Midlife ~~Crisis~~ STARTUP

Midlife ~~Crisis~~ Startup

Lessons from Venturing Out of the Ivory Tower

Lee Cooper

UCLA Anderson School of Management

NEW VENTURE PRESS
SANTA MONICA, CALIFORNIA

Library of Congress Control Number: 2003116638

Cooper, Lee G. 1944–
    Midlife ~~Crisis~~ Startup: Lessons from Venturing Out of the Ivory Tower / Lee Cooper
            p.   cm.
    Includes bibliographical references and index.
        ISBN 0-9748554-0-5 (Hardcover)
    1. New-venture initiation.   2. Entrepreneurship.
    3. Technology-enabled marketing. 4. Technology transfer.
First Edition

Copyright © 2004 by Lee Cooper

All rights reserved.
Printed in the United States of America

No part of this publication may be reproduced, stored in or introduced into a retrieval system, or transmitted, in any form, or by any means (electronic, mechanical, photocopying, recording, or otherwise), without the prior permission of the publisher. Requests for permission should be directed to:

permissions@NewVenturesPress.com

or mailed to:

Permissions
NEW VENTURE PRESS
1158 26th Street, Suite 336
Santa Monica, CA 90403

Table of Contents

Table of Contents v
List of Figures vii
List of Tables viii
Preface ix
Acknowledgements xi
PART I.   LOGIC IN USE. 1
1. Birth of a Notion 3
    1.1   Surfin' 3
    1.2   July 28, 2000 5
2. Foreplay 13
    2.1   The Conceptual Models 14
    2.2   A Dynamic Framework for Strategic Marketing Planning 22
    2.3   Into the Turbulent Field 22
3. "The Best Business Plan I Ever Read" 25
    3.1   Building a Team 40
    3.2   Dealing with UCLA 51
    3.3   The First "Public" Business Plan 55
    3.4   The First Meeting of the Board of Directors 57
4. I Should Have Read Charles Ferguson 61
    4.1   Whitewater Canoeing 61
    4.2   Return to the Real World 66
    4.3   Irrational Exuberance 72
    4.4   Due Diligence 74
    4.5   The Seeds of Conflict 77
    4.6   The Heart of Darkness 85
    4.7   All Work Is Voluntary 95
5. Smart Money 101
    5.1   The Series-A Negotiations 101
    5.2   The Prelude to Series B 105
    5.3   Don't Even Think About a Down Round 114
    5.4   Last Chance for Strategy 121
    5.5   D is for Doom 129
    5.6   Kiretsu Versus Portfolio 136
    5.7   The Tale of DVX 138

5.8		E is for Epilogue	141
PART II.		RECONSTRUCTED LOGIC	149
6.	A Linear Path		151
	6.1	Introduction to the Linear Path	151
	6.2	Kernel Analysis: Aligning Innovations with Markets	152
		6.2.1 Finding the Kernel	157
		6.2.2 Market Finding	160
	6.3	The Value of the Entrepreneurial Vision	164
	6.4	Writing a Business Plan	166
	6.5	Due Diligence on the Business Plan	189
7.	Strategic Maps		193
	7.1	Strategy as Comprehensive Problem Solving	193
	7.2	Articulating the Critical Issues	199
		7.2.1 Political Issues:	199
		7.2.2 Behavioral Issues	206
		7.2.3 Economic Issues	207
		7.2.4 Sociological Issues	209
		7.2.5 Technological Issues	209
		7.2.6 The Key Decision	210
	7.3	Mapping the Critical Issues	210
	7.4	Valuation	213
	7.5	Plans Must Be Dynamic	213
	7.6	What If?	225
	7.7	Strategic Planning Using the Four Risks	227
8.	Meta Lessons		237
	8.1	The Legend of Quincy Thomas	237
	8.2	Sharing the Map	239
	8.3	The University and Faculty Entrepreneurs	244
	8.4	The Regulating Tension of Opposites	254
	8.5	A Place to Begin and a Path to Make It Better	258
	8.6	What's Next?	259
References			261
Index			269

List of Figures

Figure 5.1. The Mental Map of Factors Affecting SDC's Success ..121
Figure 5.2. "Bake-Off Results (Static Simulator)..................................132
Figure 6.1. The Drug-Discovery Market 162
Figure 6.2. PersonalClerk's Communication with the Client Network 173
Figure 6.3. PersonalClerk Utilizes a Variety of Data Sources to Provide Real-time Marketing Messages 174
Figure 6.4. Competitive Customer Analysis Technologies. 179
Figure 6.5. Organization Chart..186
Figure 6.6. Engagement Structure 187
Figure 7.1. Critical Issues Map. 195
Figure 7.2. Potential Competitors and Features 199
Figure 7.3. The Mental Map of Factors Affecting SDC's Success ..211
Figure 7.4. Prototype for a Strategic Map 231
Figure 7.6. The Factors Impacting Market Risk.................................232
Figure 7.7. The Factors Impacting Human Risk 233
Figure 7.8. Factors Impacting Capital Risk. 234
Figure 7.9. The Complete Strategic Map for CMSS 234

List of Tables

Table 6.1. Financial Forecast Detail – Planned Forecast 184
Table 6.2. Financial Forecast Detail – Limited Forecast 185
Table 7.1. Likelihood of States in Parent Nodes. 215
Table 7.2. Two-Way Conditional Likelihoods. 216
Table 7.3. Three-Way Conditional Likelihoods. 219
Table 7.4. Four-Way Conditional Likelihoods. 220
Table 7.5. Five-Way Conditional Likelihoods. 220
Table 7.6. Six-Way Conditional Likelihoods. 223
Table 7.7. Comparative Valuations Under Different Scenarios. 226
Table 7.8. Critical Issues Facing CMSS 230

Preface

A young scientist's first lesson in scientific writing is to distinguish the flow of actions that describe the record of scientific behavior during an inquiry from the reconstruction of that record that forms the framework for a journal article. The flow of actions is called *Logic in Use* and the reorganization of that flow into the typical Introduction, Methods, Results, and Discussion breakdown is called *Reconstructed Logic*.

Part I of this book relates the flow of experiences I had mainly between the summer of 1999 and the beginning of 2002 in conceiving, creating, financing, and building a company to do technology-enabled marketing – the kind of service that personalizes the interaction of Internet users with Web merchants. Innovators, particularly university faculty or other mid-career professionals, who wish to move their ideas toward commercialization need to recognize the sometimes-subtle signs of problems or pitfalls while engaged in the incessant rush of a startup experience. But they need more than that. And thus, I have included a second part to this book that considers the steps in the startup process from a more detached, reconstructed view.

The second part deals in particular with business-plan writing and planning a business. These are not the same thing. One writes a business plan for specific reasons to specific audiences. This is the topic of Chapter 6. The resulting document is temporal and static. "Cut and paste" makes it easier to come up with the next document, but that one, too, is static. Planning a business requires a dynamic framework that addresses simultaneously the complex set of problems the business faces. Chapter 7 takes on that task, with a particular focus on university-based innovation. Chapter 8 summarizes the overarching lessons and issues I believe entrepreneurs and innovators need to consider before and during the new-venture process.

Acknowledgements

I greatly appreciate the support the Price Institute provided to develop this book and the course on "Strategic Marketing Planning for New Ventures." This support was augmented by the Price Center for Entrepreneurial Studies at UCLA and a grant from the Academic Senate at UCLA. Intel Corporation helped start me on this path with its funding (1996-99) of Project Action that allowed me to think seriously about what it means to bring radically new products to market. I wish to thank all of these funding agencies for their support. Prof. Al Osborne, then director of the Price Center for Entrepreneurial Studies at the Anderson School, UCLA, provided encouragement and support throughout this project. Thanks.

I benefited from the comments on early drafts by many colleagues at UCLA and elsewhere, particularly Steve Mayer, Marshall Goldsmith, Jack McDonough, Carolyn Cressy Wells, Ed Muller, Sam Culbert, and Alan Andreasen. Thank you for your insights and encouragement. Bill Broesamle and Gerard Rossy provided very valuable insights on drafts of the manuscript. Thanks to Fred Fox for helping me realize some of the complexities of the UC Patent Agreement and the California Labor Code. I also thank Dan Gordon for his careful reading of the manuscript, and suggestions to make it a more accessible document for its audience. I thank Jeff Marx for encouraging me to explore more deeply into the personal and emotional side of this experience, and for his valuable feedback on that side of the tale.

In Chapter 7 I have relied on parts of a planning project submitted by Ravi Narasimhan, Al Mamdani, Vijay Mididaddi, and Pak-yan (Eric) Liang for the Winter 2000 section of "Marketing Strategy in the Digital Economy." I thank them for allowing me to adapt their efforts. Ravi Narasimhan read the draft of this chapter and made thoughtful suggestions. I have used parts of a planning project concerning Core Micro-Solution Systems by Benjamin Chow, Peter Janda, Julie McDonald, Luciano Oliveira, Glenn Oyoung, and Arthur

Wang, from the Spring Quarter 2003 offering of "Strategic Marketing Planning for New Ventures." I also want to thank Sandra Fox of High-Tech Business Decisions, Inc. for providing a study of high-throughput screening. Professor C.J. Kim, Wayne Liu, and Patrick Deguzman played integral parts in helping the students and me understand the technology and preparing the strategic plan.

I want to thank each and every one of the friends and colleagues who helped me with the new venture described in this book. They took an entrepreneurial leap-of-faith with me for which I am deeply grateful. Their great skill and commitment made this the most unique adventure of my professional life. I have changed most names because, despite the personal sound and themes of the writing, it is not about them, nor is it about me. It is about recognizing, amid the rush of activities associated with any startup, the signals that say to push ahead, and the signals that say stop and think.

And finally I thank my wife, Ann, who has been there through it all, as a constant source of love, encouragement, and tolerance. I will never forget.

Lee Cooper
Santa Monica, September 2003

PART I.   LOGIC IN USE.

1. Birth of a Notion

1.1   *Surfin'*

Experienced entrepreneurs tell you to be prepared for the emotional roller coaster that launching a new venture entails. I think this evokes the wrong image. A roller coaster has a fixed track. You can see when you are climbing to the peak or diving into a valley, and you can see the bottom. You are also strapped in, and insurance companies have signed off on the risk. Your path is precisely the same as that of many who have gone before and will come after. Starting a new venture is nothing like that.

Surfing is a more apt metaphor. Before you even get wet you can watch how the waves break, look for submerged obstacles, and wax your board to minimize slipping. For the most part, you can choose your wave from how its early form looks. You can see which way the wave is breaking, and opt to go left or right on the wave. You can choose whether to kick out – if the wave walls up, cut back and let the wave reform– or ride through. The experience of a good ride on a strong wave is exhilarating, but you can also wipe out. A bad choice, or just a bad break, can send you flying into a storm of white water, crushing you under its weight – leaving you unsure of which way is up and whether you will get there in time for another breath before being sent down again. No two waves or rides are the same.

The uncertainty inherent in surfing parallels the new-venture process. Sometimes great rides are abundant and sometimes the waters are flat. Waves come in sets, as do entrepreneurial opportunities. The key is to recognize the opportunity in the early stages – when the wave is forming – ride the curl while the break is good, and kick out before the shore pound crashes you to the bottom. Easier said than done.

I hung up my homemade surfboard many years ago, when the famous storm surf of Christmas 1962 brought 30-foot waves to Malaga Cove in Palos Verdes. I watched from the bluff as veteran surfers Greg Knoll and others rode these massive forces of nature. I knew I didn't have the skill or bravura to join them. The next biggest wave I saw was almost four decades later, when the digital revolution and the Internet craze built toward a crest. I jumped on the mythical seventh wave of the seventh set. It was quite a ride. I tried to recapture the thinking that went on during the experience, the feelings both good and bad, and the emotional texture from clarity and joy to confusion and anger. For the feelings and emotions, the writing has to stand on its own. I hope I have been revealing enough to prepare you to encounter some of the vicissitudes I faced. I am eager to share the thinking that took place during the ride this book describes. I hope it aids understanding the process of new-venture initiation, particularly for university-based technology, in which radical innovations can change the way we live in and experience the world. Ultimately, this is a story about the tension between a world of technological genius and a world of business. The masters of these worlds don't know how to talk to each other. Yet, so much of the magnificent prospects for our future depend on this communication. I think managers need to extend their thinking at least enough into the technology that the basis for decision making is not opaque. And the technological geniuses of this world need to understand that, while others may be the best judges of the practicality of markets and opportunities, they more than anyone else are the best judges of the technological limits of their innovations.

I start in the middle of the ride, with the story of the first live test of the new technology, before rewinding to the beginnings of this adventure.

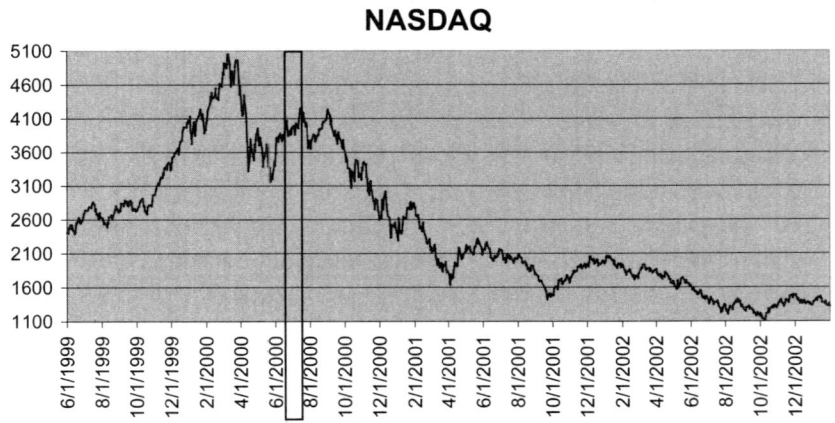

1.2 July 28, 2000

Jason Kapp picked me up at 8 a.m. from my Santa Monica home for a 9 a.m. meeting at Idealab Capital Partners, the Pasadena-based Internet incubator that spawned eToys, CarsDirect, Cooking.com, Overture, and others. Jason, our VP of client services, had played a key role in helping me start this venture. We had pitched our approach to technology-enabled marketing many times before in our successful $5 million B Round. We knew the story: The Internet dangled the prospect of huge potential returns for those who could monetize its promise for personalized shopping and browsing. But the preparation this time was different. We had heard exaggerated rumors about successes in our arena by one of Idealab's portfolio companies, as well as tales of their acumen at taking other people's ideas. So the focus centered on what to reveal and what to conceal about our approach. But the dark presence in the car, and over all discussions for almost a month, concerned when we would launch our first real market test — a test that would determine the near-term fate of this startup.

Through the B-Round representative on our board, we had arranged with iPlayer.com to purchase 20 million banner-ad impressions designated for registered users on its popular Internet versions of video games. We would use our segmentation method to learn about the preferences within each segment of its customers; we hoped to use that machine learning to increase the abysmal (and getting worse) click-through rates on the company's banner ads. Using the site's own customer data to improve ad targeting fulfilled the marketing maxim to *know your customers* and circumvented all the privacy issues concerning public-policy makers at the time. June 15 was set as the starting date. At $.75 per thousand, $15,000 to get a live test of the extension of our personalization technology in the Internet ad space seemed like a good idea. When the planned test failed to start on time, Scott Sellers, the VP for business development at iPlayer.com, told us that June inventory was sold out at $5 CPM (cost per thousand) — a much higher rate than we had negotiated. We arranged to start July 1 at $2.50 CPM, and waited…and waited some more.

Two VA-Linux machines with our software had been co-located within the test site's Web-server racks at Exodus. We had two similar machines in our own half rack at PSINet in Marina del Rey, and four more tied to a T-1 line at our office in Santa Monica. Tests on our end were fine so far, but within iPlayer.com's network we couldn't read the cookie — that tiny piece of text code that gives the originating site so much information about the customer at the other end. We needed the site to change six lines of code to share cookies just within its complete subnet. Since my technical expertise covered not cookie logic but the conceptual and analytical models that drove our learning and optimization algorithms, I was barely grasping onto these problem areas. Early in July, the client's CTO promised to make the simple change we required, but it didn't happen. On July 19, the CTO left for a year, supposedly to recover from an unspecified illness. I suspected burnout. It was July 25 before Sellers got his team to put the new cookie code in place. We found and fixed a small bug in our database agent, and sent a ready-to-go message to Sellers. Ravi Srinivasan, the head of our technical-implementation team and an Anderson MBA student, heard that AdForce, the firm serving iPlayer.com's ads at that time, had been notified to start sending our ads first thing on July 27…and still we waited.

Jason and I agreed to show Idealab much the same demo that had been developed in record time for our first board of directors meeting in early February. The demo featured our product-recommendation suite, called PersonalClerk, that incorporated the Internet advertising optimization only as a simple, natural extension of the personalization solution. We wouldn't talk about the ad test with Idealab.

The 45-minute meeting went nowhere, but went nowhere smoothly. We learned Idealab's efforts were mostly vapor — great customer data, but no sense that the company knew how to use it effectively. Technology-enabled marketing operates at the boundary between intelligent information systems and that obscure area called *marketing science*. Without strong capabilities in both areas, the problems are at best half solved and the solutions, consequently, are half-baked.

Once back in the car I called Ravi. He sadly reported, "No data yet." Then I heard Chuck Yu, our hardware guru, in the background yelling that the first ads were being sent. It was 9:52 a.m. and our labor pains had just begun. The drive back felt like a rush to the hospital to be in time for the delivery of my first child.

When Jason and I arrived back at the bullpen on the second floor, where most of the technology group sat, everyone was gathered around Chuck's desk. One window on his Linux notebook tracked the number of open sockets – one for each active user. The theory was that when a user's browser requested a page, our HTTP agent that handled the raw traffic opened a socket as a file description, delivered an ad, sent the needed information to our learning algorithm and a session log, and closed the socket. We watched as the number of open sockets increased toward 1,024, blocking the thread and crashing our system. Our design didn't threaten to bring down the client site or interfere with its basic operation, but when our system crashed, the ads we paid for weren't being delivered. The data that were crucial to our machine-learning algorithms weren't being collected. Chuck's fingers flew over the keyboard each time the thread was blocked. If the sockets were unblocked (closed) when the process crashed, a simple chain of commands would restart the process immediately. If the sockets remained open – blocked from accepting a new-user request -- Chuck had to reboot our Linux box inside the client's rack at Exodus before restarting the process. Then learning would begin again, until the next blockage.

Why was this happening? What could we do about it? I didn't have a clue. Chuck was madly typing away to keep our downtime to a minimum, and I had no insight into the problems we faced. Thirty years of designing and building medium- and large-scale analytical and statistical systems for squeezing meaning out of market data, and I had never been this clueless. Others had done major chunks of many of my earlier projects, sometimes for efficiency and sometimes for learning. When stumbling blocks were encountered, I had always been there to solve the problem. Not this time. After more than three decades on the faculty of UCLA's management school, I was now outside the ivory tower. I needed to step back, micro-manage less, and grant other people control over a problem-solving process that transcended my expertise.

The intellectual resources available were substantial, but incomplete. Giovanni Giuffrida, a UCLA doctoral candidate in intelligent information systems, was our CTO, and had developed wonderfully into a leader of the technology group. An ever-reshuffling handful of program developers would gather around the conference table in his office next to the bullpen while Giovanni worked with them through whatever was the greatest barrier to our technical progress. Giovanni and I previously had worked together for several years on large-scale forecasting projects[1] and research-oriented datamining projects that had led to the first datamining article published in the mainstream management literature.[2] Our big-data experience mostly concerned retail scanner records. We had to deal with about 25 million records at a time and be prepared to create up to 800 million forecasts a year. Those are very small numbers compared to what we faced on the Web.

Our main Web expert, Fabrizio diMauro, an amazing code hacker, was stuck on a trans-Atlantic flight returning from Italy – and grounded in Newfoundland when his girlfriend became extremely ill in flight. When Giovanni, Jason, David VanArsdale (VP/admin) and I had started this venture together, Fabrizio was the first person hired in the technology group. Giuseppe Blanco, a Web-design specialist, C programmer, and the third part of the Sicilian Connection, was in the bullpen. Not only did these three grow up in the same small town in Sicily, they all reconnected in computer-science graduate programs at UCLA. Brandon Davinski, the UCLA computer-science undergrad who worked with us full time in the summer, brought an almost scary knowledge of Web programming. He already had found a major security hole in another client's cash-register program. He could change the price for anything in his or anybody's current shopping basket. Many times I mumbled to myself that I was glad Davinski was on our side. Wesley Rhim was a database expert with deep SQL and Perl skills. Murilo, a great C programmer with a PhD in physics, had just started earlier that week. Nick, a computer-science undergrad from MIT, worked with us that summer. So did Jonathan, the best computer mind in my older son's cohort (then 20), who had programmed the common object module (COM) needed for dealing with the Microsoft servers that the client used. In supporting roles were a group of very talented college students in their first important summer jobs.

At first, the core technology team tried to force the sockets to close – fix the symptom and ignore the problem. We didn't find a way to do this. Maybe the level of simultaneous traffic was too heavy for only 1,024 sockets, but this seemed unlikely given the short duration

---

[1]Cooper, Lee G., Penny Baron, Wayne Levy, Michael Swisher, and Paris Gogos (1999), "PromoCast: A New Forecasting Method for Promotion Planning," *Marketing Science*, 18, 3, 301-316.
[2]Cooper, Lee G. and Giovanni Giuffrida (2000), "Turning Datamining into a Management Science Tool," *Management Science*, 46, 2 (February), 249-264.

(milliseconds) that each request required. So Giuseppe and Jonathan pulled a 10-thread version of the HTTP agent from the program repository and tested it. Chuck copied it over to the machines at Exodus, compiled it – keeping the other system going to the last second -- and then started the 10-thread code. The inexorable rise in blocked sockets foretold the outcome. Blocked sockets led to crashed threads. It took 10 times as long to crash the whole program, but this try eliminated traffic volume as the possible cause. To eliminate bandwidth issues, the banner ads were copied over to the extra machines at PSINet and served from there. No discernible difference.

By mid-afternoon, someone suggested that possibly old or strange browsers contributed to the problem. Our session logs were full of requests from newer Internet Explorer and Netscape browsers, but none from the much older versions of these or from WebTV browsers. Jonathan grabbed a WebTV emulator that we used to log into iPlayer.com. No ad was delivered and no record of the user showed up in the session log. This signaled that the problem was in how our HTTP agent handled old or odd browsers. The HTTP agent was our interface to Internet traffic – in essence, a Web-server operating system stripped down for speed. We could handle about 1,200 requests per second, per box in a system we knew how to easily parallelize, if additional speed was needed for high-volume commercial sites. But apparently, too much of the browser handling had been stripped out. The fix required putting a whole APACHE Web-server operating system in front of our HTTP agent to take over the browser handling. Giovanni understood the best talents of his core team and assigned David and Jonathan to the APACHE tasks, Giuseppe to the modification of the HTTP agent, and Nicholas to the required CGI scripts. Chuck was still typing rapidly and constantly to keep the processes almost continuously available. Murilo poured over APACHE support documentation, regretting that as a "newbie" he couldn't be more central to the excitement of the problem-solving process. His time would come.

My role shifted from quietly making sure that the discussions, diagnoses and possible remedies made sense to even more quietly making sure the group had enough pizza and sodas to keep them working. As long as I stayed there, I knew they would stay. That much mutual respect we had built in the eight months of working together on the multitude of technology problems startups must solve to endure. Unfortunately, the office space we leased wasn't set up for the long hours of startup companies. The air conditioning shut off at 7 p.m. on that hot July Friday. I had the unassigned troops getting all the fans from our 6,000 square feet of offices into the technology bullpen and the corridor outside in order to usher the cooling evening air into an overcrowded work space.

Close to 10:30 p.m., the pieces began to come together. Chuck started sending chunks of code over to Exodus and bolting them in place: copy, compile, start, check the logs, try the WebTV browser, and check the logs again. The system held. By 11:10 p.m. we knew we had succeeded. Chuck had spent more than 13 hours keeping our baby alive as it worked its way through the birth canal. We all felt like proud new parents; the sense of accomplishment was palpable. We made bets on what final click rates our optimization would ultimately achieve. Chuck set up the monitors that would ring the cell phone and send emails if the baby sneezed, and we all prepared to go home – having wasted fewer than 150,000 ads from the 20 million we had purchased.

The bonding that went on that day-turned-night had a lasting effect on the performance of the technology group members. Their individual skills in problem diagnosis and code creation proved instrumental to achieving a desired outcome. They all felt a deep stake in the performance of the software and the company. In important ways, their work became a mission rather than merely a job. The focus, tenacity and dedication they continued to exhibit, I attribute in no small degree to what we all experienced that night.

Something special happens when people are connected to the mission of the organization, value what the company wants to accomplish, and sense an alignment between their skills and efforts and the sought-after organizational goals. I learned this lesson 25 years ago when I directed the UCLA Arts Management Program – an MBA program that trained managers primarily for not-for-profit performing-arts organizations, museums and arts councils. The mission of a not-for-profit arts organization has value – both conceptual value and monetary value. The monetary value is easy to see when people volunteer their discretionary time and money to help the ballet or opera, or when others work for much less compensation because they are "working in the arts." The conceptual value is less tangible. Caring about the organization is a partial antidote to the bureaucratic barriers and careerist silos that trap so many other organizations. While not a cure for ineptitude, the conceptual value of an organization's mission helps control behavior within the organization.

I once thought these organizational advantages accrued only to the not-for-profit sector.[3] My experience in founding Strategic Decision Corp. taught me, however, that entrepreneurs have the potential for gaining much of the same leverage. One of my goals for this book is to share my experiences and offer perspectives that can help entrepreneurs, practicing managers, and management students build successful organizations in which people can invest their hearts along with their minds. Another goal is to aid entrepreneurs inside and outside the ivory tower who need to understand the venture-initiation process along with the opportunities and traps that lie in wait. A final goal involves describing a modern approach to strategic marketing planning for new ventures. Along the way I will suggest how the mantra of *segmenting, targeting* and *positioning* that has characterized marketing education for the last 25 years can be updated to reflect the realities of technology-enabled marketing in the digital world.

---

[3]Cooper, Lee G. "Some Perspectives on Art, Organizational Behavior and Democracy," *The Journal of Management and Law of the Arts*, 11, 1-26 (1981). Perloff, Harvey S., Paul Bullock, Lee G. Cooper, Simon Eisner, and Hyman R. Faine (1979), *Arts in the Economic Life of the City*, New York: American Council for the Arts.

2. Foreplay

This chapter steps back and develops the thoughts and theories that shaped my approach to this new venture. The story line traces the origins of my Intel grant and how obtaining it led to my commitment to build a company.

I peered through the candy-store window in the 1970s while my lifelong friend Steve Mayer helped create the video-game industry as one of the pioneers of Atari. My diverse palette of marketing-research techniques offered little value as his creative instincts and engineering acumen shaped the tastes of a generation. Simply putting a new game in a few arcades and counting the quarters at the end of the week passed for research in those early days. The best games stopped working because the coin boxes were jammed to overflowing. I got tenure and he became rich and famous – both welcome outcomes from my point of view. The desire to shape my expertise to his domain, however, was never sated.

Marketing research does its proper job assessing consumer response for either existing products or new products in existing markets. The research basis for creating new markets is highly uncertain, but has piqued my interest over the years. My chance to work in this arena came when Martin Greenberger, the IBM chaired professor and director of Anderson School's Center for Digital Media, asked me to accompany him to the Intel Architecture Labs in Oregon in the mid-'90s. The plan involved spending Monday listening to Intel teams report on research projects the company was undertaking, and then on Friday, having Intel representatives meet us at UCLA to listen to our research ideas. If common interests emerged, Intel would fund them for a year as an experiment. Just listening to their half-dozen extended reports gave us more insight into how they view the world and what they worried about. Of the many projects each of our team members had going on, we picked a small set of those that seemed best aligned with Intel's efforts. Several projects were approved for first-year (of possibly three years) funding. My effort, named "Project

Action" by Martin as a follow-up to his "Project Vision," was designed to address the issues of doing marketing research on radically new products – products that fundamentally change the way consumers think about a domain or what they expect from products in that domain. I received funding for three years and developed an approach to strategic marketing planning for such products.

2.1 The Conceptual Models

The grant, along with a sabbatical year, gave me the opportunity to study the radical-innovation process. The remainder of this section, along with the next, sketch the conceptual models and theories that I assimilated and used in my approach to planning for and managing the process of bringing radically new products to market.

I learned from Stuart Kauffman that in periods of radical change we expect grossly different forms to evolve to serve basically similar functions.[4] The Cambrian explosion heralded the creation of 100 new phyla – only a third of which survive today. Instead of the "gradual accumulation of ... profitable variation" – the Darwinian notion of evolution by natural selection[5] – the Cambrian explosion leads us to expect long leaps across ecological landscapes. In our applied context we wonder what happens when the wild flight of entrepreneurial vision and creativity hits ground. In short, the local business ecosystem must be rich enough to nurture the innovation or it will die.

The biological flavor of this is not accidental. I very much see organizations as living entities attempting to navigate a path through a mixed economy. Like all living organisms, organizations have semi-permeable boundaries separating them from the environment. Resources flow both ways across organizational boundaries in what must be a long-term profitable exchange with the environment for the organization to survive. I look at marketing as a boundary-management function – controlling the flow of resources across organizational boundaries. If we think of marketing communications

---

[4] Kauffman, Stuart E. (1995), *At Home in the Universe: The Search for Laws of Self-Organization and Complexity*, Oxford: Oxford University Press

[5] "No complex instinct can possibly be produced through natural selection, except by the slow and gradual accumulation of numerous, slight, yet profitable, variations." Darwin, Charles (1859), *On the Origin of Species by Means of Natural Selection*, p. 210.

as the messages an organization sends across its boundary into the environment, the analogy seems apt. Managing the image of a company, the face it puts to the outside world, is indeed boundary management. Certainly, product-management issues such as determining feature sets, price, promotion policy and distribution policy are the traditional concerns of how companies market their offers to consumers. I believe it is still apt to think this way when the subject is how a business markets itself to the financial community. CEOs will tell you communicating with Wall Street is very much part of their ongoing marketing efforts. Similarly, when companies seek the specialized labor needed to develop products, they are marketing the company to the labor pool. And marketing has a major role in facilitating the alliances and partnerships that companies form to ensure that a whole-product solution is offered. Consumer segments, the investment community, labor markets and relevant business partners are interdependent outside environments that an organization must successfully navigate to survive. While marketing communications is often the only formal function of a dedicated marketing department, this is due to an expanded role for marketing at the top levels of the firm.

This living-systems view of organizations is one I've held since my exposure in the 1960s to general-systems theory.[6] My first important job was for the Advanced Marine Technology Division of Litton Industries. This hot, naval-systems design team was in the middle of winning a series of government ship-design contracts totaling over $4 billion. Nepotism created an opportunity in the summer after my first year in grad school. My father was on the senior scientific staff and summer openings were made available to college-age sons and daughters. While others worked in the library, mailroom and clerical positions, my credentials (and my father's reputation as an extraordinary problem solver) landed me a job in personnel subsystems design. Function and task analysis, systems operability and maintainability, and human-factors engineering were all tools and concepts that enabled the design work. Underlying them all was living-systems theory. The DOD-approved design approach was called Design-Work Study. But my father advised me to use the underlying theory and my own problem-solving skills, then

---

[6] See von Bertalanffy, Ludwig and Anatol Rapoport (1956), *General Systems: Yearbook of the Society for the Advancement of General Systems Theory, Volume 1*, Ann Arbor, MI: Society for General Systems Research.

reconstruct the solution into the approved language. As a result, thinking in living-systems terms became second nature for me.

Accelerating computer capabilities enabled management scientists and operations researchers to cast many systems problems as large-scale numerical-optimization algorithms. The use of general systems theory waned in the 1970s and 1980s as it resisted mathematical formalization. But the more recent rise of agent-based approaches, including genetic algorithms, genetic programs, artificial life, and complex adaptive systems, helped bring old thoughts back into vogue. So, while more than half a century old, living-systems thinking is very consistent with the modern management thinking of James Moore[7] when talking about how traditional notions of competition are antiquated, and Clayton Christensen[8] when very directly discussing the forces that control the fate of disruptive technologies. Christensen, extending the theme earlier developed by Richard Foster,[9] asserts that radical innovations first appear as inferior goods. Steamships, for example, initially seemed no threat to the clipper-ship franchise on trans-Atlantic freight. Steamships were more expensive per ton, delivered smaller payloads, and were initially less reliable. So the freight companies listened to the voices of their best customers and increased cost efficiency by allocating corporate resources to craft clipper ships with more masts, sails and cargo capacity. Of course, these freight companies perished when the steamers emerged from the more protected environment of river shipping, which fundamentally valued the steamship's core advantage (the ability to navigate regardless of the prevailing winds) and did not need what steamers couldn't initially provide (the large payload capacity and reliability required for trans-Atlantic efficiency). In 1783, the steamboat *Pyroscaphe*, built by the Marquis de Jouffroy d'Abbans, was tried out on the River Saone. In 1802, Symington's *Charlotte Dundas* was used as a tugboat on the Forth-Clyde canal. The next year American Robert Fulton demonstrated a steam-driven boat on the

---

[7] Moore, James F. (1996), *The Death of Competition: Leadership & Strategy in the Age of Business Ecosystems*, New York: Harper Business.
[8] Christensen, Clayton M. (1997), *The Innovators Dilemma: When New Technologies Cause Great Firms to Fail*, Boston: Harvard Business School Press. Bower, Joseph L. and Clayton M. Christensen (1995), "Disruptive Technologies: Catching the Wave," *Harvard Business Review*, January-February, 44-53.
[9] Foster, Richard (1986), *Innovation: The Attackers Advantage*. New York: Simon & Schuster.

River Seine.[10] Given that steamships found a friendly, nurturing environment, they evolved by incremental improvements until they became more-than-viable competitors to the clipper ships. By 1820, 35 paddle steamers were in regular use on the Mississippi. Only a year earlier had the *Savannah* become the first steamship to cross the Atlantic. It took 25 more years before Cunard's *Hermann* and *Washington* steamships provided a regular cross-Atlantic service. This is more than just one technology supplanting another. The *companies* that dominated the freight trade using clipper ships failed. Their planning processes always favored allocating resources to sustained, incremental innovations on clipper ships rather than to nascent, competitive technology such as steam. By the time they saw the consequences of their strategy, it was too late. The innovative technology was on a technology-improvement curve with a much steeper slope. The clipper ships were about to be irrevocably eclipsed. This is what Foster meant by "…The Attacker's Advantage."

From the point of view of the attacker, Christensen asserts, innovating companies face a fundamental choice when commercializing a disruptive technology. Option 1 is to accept the market's needs as well defined and push the technology to its limit in addressing those needs. Option 2 is to accept the technology's current capabilities as a given and seek the market that will value the inherent attributes of that technology. Christensen asserts, and I agree, that Option 2 is the more successful route. This is a lesson to which we will return later: Find the kernel of innovation and understand its capabilities and limitations. Understand that asking the standard questions of the best customers can be extremely misleading in an emerging market environment. *Market finding* is required. Research that seeks the right market, business ecosystem, or value network may be better than research that tries to tailor product attributes as if they were a sustaining technology.

The other big lesson concerns the change in industry structure described by many current authors, but which has its roots in the classic work of Emery and Trist (1965).[11] Andrew Grove[12] recounts

---

[10] See http://www.saburchill.com/history/events/024.html for a timeline of the steamship evolution.
[11] Emery, Fred E. and Eric L. Trist (1965), "The Causal Texture of Organizational Environments," *Human Relations*, 18, 1, 21–32.

the structure of the computer industry in the 1960s, when IBM dominated the mainframe arena. IBM developed its own CPUs, memory and storage technology, operating systems, applications software, manufacturing, sales and distribution, service, and training. DEC performed the analogous functions in the minicomputer market. In both vertical slices, large organizations allocated resources to internal development according to the organizations' sense of company priorities – possibly listening to the voice of their best customers. Initially this might have been the most efficient approach, given the paucity of the necessary problem-solving skills in the general environment. But as the sector matured and universities prepared more and more engineers and physicists, smaller specialized firms began to emerge.

Coase's thinking on firm size and transaction costs explains some of the dynamic.[13] He asserts a firm will tend to expand to the point where "the costs of organizing an extra transaction within the firm become equal to the costs of carrying out the same transaction by means of an exchange on the open market." So the vertical dinosaurs ruled the computer landscape when the expertise was narrowly held. The search costs to find buyers and sellers were huge, as were information costs, bargaining costs, decision costs and enforcement cost.[14] Optimal firm size was understandably large. But to maintain the dinosaur status once expertise was more widely available, IBM had to be nearly the best of breed in all the separate functions. The downsizing and outsourcing trend of the 1980s accelerated a perhaps inevitable process by ensuring a ready supply of experts and innovators to compete for each element in the value chain. As the transaction costs drop, the optimal firm size drops. In the digital economy, transaction costs are dropping toward zero, with startling implications for optimal firm size. We should not, then, be surprised that providing a whole product in high-tech arenas takes a network of original equipment manufacturers, operating system vendors, independent hardware vendors, independent software vendors,

---

[12] Grove, Andrew S. (1996), *Only the Paranoid Survive: How to Exploit the Crisis Points That Challenge Every Company*, New York: Doubleday.
[13] Coase, Ronald H. (1937/1952), "The Nature of the Firm," in *Readings in Price Theory*, George. J. Stigler and Kenneth. E. Boulding, eds. Chicago: Irwin, 331-51.
[14] Cf. Shapiro, Carl and Hal R. Varian (1999), *Information Rules: A Strategic Guide to the Network Economy*, Boston: Harvard Business School Press. Robertson, Thomas S. and Hubert Gatignon (1998), "Technology Development Mode: A Transaction Cost Conceptualization," *Strategic Management Journal*, 19, 6, 515-531.

systems integrators, distributors, trainers, and service organizations – smaller organizations whose fates are basically correlated.

The evolution of smaller, specialized firms is what Emery and Trist (1965) expect in times of turbulent change. They describe *Turbulent Fields:* "In [turbulent fields], dynamic processes, which create significant variances for the component organizations, arise from the field itself. The 'ground' is in motion" (p. 26). In an environment with this much uncertainty, Emery and Trist believe that certain social values will emerge as coping mechanisms. To succeed in this environment, an organization must form organizational matrices or "relationships between dissimilar organizations whose fates are, basically, positively correlated" (p. 29). An organization must also strive for institutional success by working toward goals that fit its character and by moving in a direction that converges with the interests of other organizations in the matrix.

In times of turbulent change, the formerly static, vertical industry structure is supplanted by webs of mutually dependent economic interests. Such economic webs are discussed directly by Kauffman (1988),[15] Hagel (1996),[16] James Moore (1996),[17] and more generally by Geoffrey Moore (1991, 1995)[18] in his descriptions of how, in order to *cross the chasm,* firms must partner to ensure that they deliver a whole-product solution that provides the economic buyer a compelling reason to buy.

I felt Moore's (1995) work, in particular, gave a sound basis for adapting these open-systems principles into high-technology contexts. Moore discussed his version of the technology adoption life cycle (TALC) as a special case of the product life cycle taught in intro marketing classes. In *The Early Market* technology enthusiasts are captivated by the exciting new possibilities an innovation presents.

---

[15] Kauffman, Stuart E. (1988), "The Evolution of Economic Webs," in *The Economy as an Evolving Complex System, SFI Studies in the Sciences of Complexity*, Philip W. Anderson, Kenneth J. Arrow, and David Pines eds., Reading, MA: Addison-Wesley Publishing Company.
[16] Hagel III, John. "Spider versus spider." *The McKinsey Quarterly*, 1 (1996).
[17] Moore, James F. (1996), *The Death of Competition: Leadership & Strategy in the Age of Business Ecosystems*, New York: Harper Business.
[18] Moore, Geoffrey A. (1991), *Crossing The Chasm: Marketing And Selling Technology Products To Mainstream Customers*, New York: Harper Business. Moore, Geoffrey A. (1995), *Inside the Tornado: Marketing Strategies from Silicon Valley's Cutting Edge*, New York: Harper Business.

They patch together working systems from incomplete components. Business visionaries who see the possibility of first-mover advantages supplant the enthusiasts and help shape the nascent technology to serve business needs. During *The Chasm* (the topic of Moore's 1991 book) early enthusiasm wanes and potential mainstream customers take the show-me attitude of economic buyers – requiring a whole-product solution and a compelling reason to buy. Across the chasm is *The Bowling Alley* – a niche-based opportunity. The innovating company survives by crafting a whole-product solution that provides the economic decision maker in a particular vertical market with a compelling reason to buy. This represents the first pin in the bowling alley. The bowling-alley strategy progresses by extending the initial application to a second vertical segment, then a third segment, and on, or providing a second application to the initial vertical segment, then a third application to this initial segment, and on. By the end of this process, the innovating company has enough segments and applications of its core technology deployed to be a candidate for becoming a required part of the infrastructure of the whole business ecosystem. If the business ecosystem adopts the technology as a standard part of business, and the ecosystem goes into hypergrowth, then the fate of your company is tied to your ability to supply the needs of the rapidly growing category. *The Tornado* reflects the period of mass-market adoption of the innovating company's core technology. *Main Street* comes after the period of infrastructure deployment. The concerns in this period are much the same as those of most of modern brand management in mature categories – developing brand extensions and product-line extensions to create opportunities within particular market segments. *End of Life* is the period when wholly new disruptive innovations threaten to overturn and supplant the infrastructure of the existing business ecosystem.

While simply articulating the stages can help shape the managerial thinking of entrepreneurs trying to bring radically new technologies to market, the implications for strategic thinking are in some ways more profound. Living in mainstream markets has led marketing and organizational scholars to think of the goal of corporate strategy as building sustainable, competitive advantage. Focusing on these six stages as the expected consequences of the technology-diffusion process might help strategic thinkers realize that no static plan can provide sustainable advantage.

Shifts in strategy required for a company to remain successful through the stages of the TALC. For example, how you deploy your sales force to target buyers shifts from stage to stage. In *The Bowling Alley* you focus on the economic buyer and the end user – approaching the infrastructure buyer late in the sales cycle. In *The Tornado* you ignore the economic buyer and the end user – focusing exclusively on the infrastructure buyer. In *Main Street* you sell to the end user. The messages you send to buyers also shift. In *The Bowling Alley* you emphasize return on investment as the compelling reason to buy. In *The Tornado* you ignore return on investment – focusing instead on timely deployment of reliable infrastructure. In *Main Street* you focus on end users' experience of the product, seeking to gratify their individual needs.

Regarding product differentiation, in *The Bowling Alley* you differentiate your whole product for a single application. In *The Tornado* you commoditize your whole product for general-purpose use. In *Main Street* you differentiate the commoditized whole product with specialized marketing campaigns with product features targeted toward specific market niches. Distribution strategy also changes predictably. In *The Bowling Alley* you partner with a value-added distribution channel to ensure customized solution delivery. In *The Tornado* you distribute through low-cost, high-volume channels to ensure maximum market exposure. In *Main Street* you continue to distribute through the same channels, but now focus on merchandising to communicate an extra feature that appeals to each main-street market niche. Pricing strategy changes. In *The Bowling Alley* you use value-based pricing to maximize profit margins. In *The Tornado* you use competition-based pricing to maximize market share. In *Main Street* you celebrate that extra feature for each niche to gain margins above the low-cost clone. Competitive reaction changes. In *The Bowling Alley* you avoid competition to gain niche market share. In *The Tornado* you attack competition to gain mass market share. In *Main Street* you compete against your own low-cost offering to gain margin share. And finally, the markets you target change. In *The Bowling Alley* you position your products within vertical market segments. In *The Tornado* you position your products horizontally as global infrastructure. In *Main Street* you position yourself in niche markets, based on the individual preferences of end users.

The gestalt that emerges from the pattern of strategic shifts in this exhibit is of an innovative company as a living organism – constantly sensing the nature of the environment or ecosystem it is in or about to enter, adjusting its boundaries to sense what is needed for survival and growth, and responding by making internal adjustments to provide what the local environment demands and (given that their fates are positively correlated) forming the alliances with other living systems that are needed to confront the anticipated obstacles. This is how living systems confront a turbulent field.

2.2 *A Dynamic Framework for Strategic Marketing Planning*

If the world around our enterprise is constantly changing, what good is strategic planning? I believe that strategic planning is essential; the problem is that, once written, strategic plans are instantly out of date. Planning should compel managers to confront the rich complexity of interrelated issues facing their venture. Putting that understanding into a written document freezes the result in time, while time has the inevitable property of moving on. What is needed is an approach that captures the knowledge and understanding of the problems, but is as dynamic as the changing times demand. Chapter 7 deals with this topic in detail. Here I'll just mention the steps:[19] Find the business kernel of the innovation; identify the appropriate first market; articulate the value proposition for that market; list the stakeholders; sketch the venture in value-network terms; specify the critical issues facing the venture, the business sector, and the larger infrastructure surrounding the sector; and map the issues into a strategic map or networks that allows you to run the scenarios for best case, worst case, expected case, and other interesting cases. An example of this approach as applied to Strategic Decision Corp. is given on the VentureDevelopmentProject.com Website.

2.3 *Into the Turbulent Field*

Armed with these ideas and frameworks I began teaching product strategy in the digital economy in early 1999. I had gone halfway the year before, teaching a course split between radically new products and efficient consumer response (ECR). ECR was the largely Procter

---

[19] Cooper Lee G. (2000), "Strategic Marketing Planning for Radically New Products," *Journal of Marketing*, 64, 1 (January), 1-16. Cooper, Lee, Troy Noble, and Elizabeth Korb (1999), "Strategic Marketing Planning in Turbulent Environments: the Case of PromoCast," *Canadian Journal of Marketing Research*, 18, 46-66.

& Gamble-backed industry initiative to get the right product to the right place at the right time in the right quantity, assortment, and price to meet consumer needs. While the concept was intriguing (i.e., to use detailed demand indicators at the retail-store level to drive decision making throughout the supply chain), teaching it was extremely boring – particularly when juxtaposed with the excitement around radical innovations. So I dropped ECR from my next version of the class, concentrating on Geoffrey Moore's (1995) framework for the technology-adoption life cycle.

This new course attracted a group of entrepreneurially oriented MBA students – caught up in the Internet fever of the time. Student teams did strategic-marketing-planning projects for Scour.net, electric vehicles, Swatch® Access (a smart-card technology integrated into watches), enhanced TV, video on demand, Internet shopping bots, National Semiconductor Corp.'s PC-On-a-Chip, Power Pay (an Internet payment method), Adaptec's Satellite Express® (a direct satellite to PC connection), and Olean® (Procter & Gamble's fat substitute).

These students were eager and focused learners who took initiative in their own education. I hadn't seen such motivation since the Arts Management students I'd taught in the '70s. I told myself that I just wanted to understand better the point of view of these entrepreneurial students. So by June, I had signed up for the September offering of Guy Kawasaki's Garage.com *Bootcamp for Startups*.

I also worked that year with the Field Study Program for the Executive MBA students. The 72 students in the EMBA Program formed a dozen teams, with each team taking on two projects. The first was focused internally on some issue important to the host company – Sun Microsystems in 1999 – and the second was focused on an issue important to one of Sun's major customers. Each faculty member advised two of the teams. My teams were dealing with projects for Daimler–Chrysler and Qwest in addition to their internal projects for Sun. Tackling real projects for real companies was fun. It also provided a way to relate to MBA students on a much more even and interactive basis than I could when teaching advanced statistics and marketing-research methods to students who, frankly, didn't care. These studies always ended with the host company taking the students, faculty, and the host's clients that were part of the study, off to a resort/conference center for final reports. In 1999, Sun took us all to Sophia Antipoli, the industrial park/conference center in the South of France near Cannes.

I felt attracted to the growing excitement surrounding technology markets and the Internet. But I did not yet know that I could have any role other than consultant or faculty supervisor. Most of my established skills were in the advanced-analytical-modeling arena. But events in this world seemed to move too fast for the kind of study that I historically undertook. In late July of 1999, listening to our Executive MBAs present final reports to Sun Microsystems on their half-year-long studies, ideas began to brew. One evening, walking with my wife along the shore, it came to me. I snapped out of my silent reverie to announce that I just had a great idea. Ann and I often walk. Watching me drift off from conversation into some remote, private arena and then suddenly return to the present wasn't a unique experience for her. Perhaps she sensed that this time the focus was different. I spoke of business ideas, not research. I said I could use the datamining methods I'd been working on with Giovanni to do real-time, technology-enabled marketing – not to study online markets, but to be part of the online enterprise: segmenting, targeting and recommending based on the empirical record of online customer behavior.

I spoke about this idea with two of the graduating EMBA students who were with me in France. David VanArsdale was deputy director of the Anderson Computing Center, having come years before to help design and oversee the implementation of the then-state-of-the-art computing network and facility for the new Anderson School complex. With that task accomplished, he had enrolled in the EMBA Program to prepare for the next steps in his career. Teri Connelly had a background in marketing, including datamining in the telecommunications and utility arenas. Both were interested and at a point in their lives when such a new venture might be worth the risk. In my mind, I had committed to building a company.

3. "The Best Business Plan I Ever Read"

Here I tell the story of the creation of Strategic Decision Corp. and show the development of its basic business plan, from the earliest versions to the version presented at the first Board of Directors meeting that garnered the praise in the chapter title, from billionaire Fred Hart.

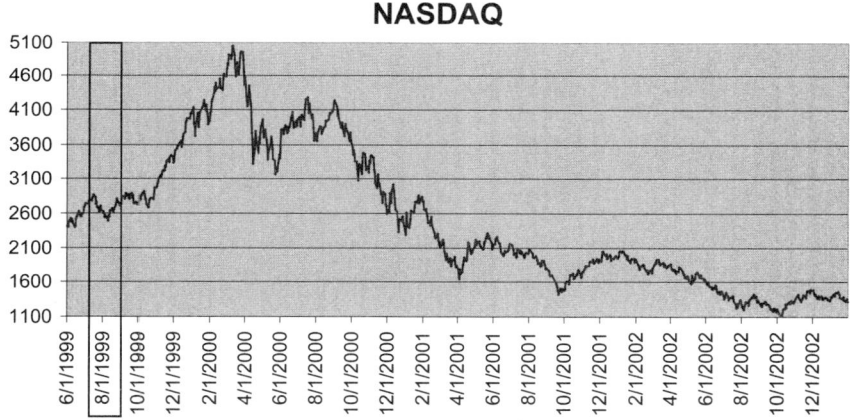

Call Giovanni. That was my first task upon returning from France. I knew he had a venturesome side. We had previously discussed possible business connections even after our project with *ems, inc.* was completed. As our datamining article wound its way through the editorial maze at *Management Science*, Giovanni took a practicum job at Hughes Research Labs (HRL) in Malibu and worked during off hours on his dissertation.

My pitch to Giovanni was essentially this: Privacy was a large and growing concern in public-policy circles. This concern would most likely block any trans-site customer-intelligence utilization, largely neutralizing the incumbent advantage of companies such as Doubleclick or Engage that could follow users around the Internet. But the basic marketing tenet of "know your customer" ensured the value of using site-specific customer databases to make more appealing offers to e-commerce sites' registered users. When *your* customers show up in *your* store, serving them better with the help of data they provided is simply smart business, not a violation in any sense of basic privacy. For an e-commerce site, datamining could personalize the experience like a *personal clerk* – silent and unassuming but aware of a customer's tastes and spending patterns – never forgetting his or her name, past purchases, preferences, or complaints. This approach fit well with two technical aspects of the datamining algorithms Giovanni developed. First, the rule-discovery system was written as a superset of SQL (Structured Query Language). This meant that the customer data that resided in large corporate databases never had to be moved outside the protection of the corporate firewall. The SQL instructions for rule discovery could be sent (even across the firewall if necessary) to the database, and the computer and database program that were already in place could do the heavy lifting. A nice side benefit for us was that our cost structure for equipment could always be tiny compared to what the client company was already spending on its IT infrastructure. Second, the datamining algorithm allowed for incremental knowledge acquisition. Thus, if we datamined a 25-million-record database that grew by, say, 300,000 records a week, we could mine just the 300,000 new records and add them to the already-established knowledge base. (Just a few months later, during discussions with IBM on possible partnering, the company would question why anyone would want such capability when, with enough of its hardware, IBM could completely re-mine an 18-terabyte database overnight. Giovanni and I just looked at each other and covertly smiled.)

Of course Giovanni was interested in my idea. His role was obvious and very well aligned with his skills and interests. It involved no major risks for him – he could continue at HRL, just as Van could continue at Anderson, Teri could pursue her job search, and I could continue at UCLA – as plans became more concrete.

I attended Garage.com's *Bootcamp for Startups* in mid-September (remember, I registered for this in June under the pretext of getting closer to the mind-set of my entrepreneurial students). Guy Kawasaki was at his evangelical best, presenting an impressive collection of top speakers and panels. The audience enthusiastically consumed every tidbit of insider knowledge. After one panel consisting of venture capitalists and angel investors, Guy provided the opportunity for the first five attendees getting to the stage to give a 45-second pitch and receive an immediate thumbs-up or thumbs-down from each panelist. I stared in wonder and amazement as one of my former students, Al Shriver, jumped to the stage. I don't remember his idea — just that he sailed into the pitch and was undaunted by the thumbs-down. In my class, Al had done a strategic marketing plan for an Internet payment scheme. The resulting analysis convinced him it wouldn't work — saving him whatever effort he might otherwise have spent on that idea. This experience convinced me of two things: First, by Al's 10th idea he will be a great success. Second, this could not be my approach to entrepreneurship. In my mid-50s I'm not a jump-on-the-stage kind of guy. Actually, I never was.

The atmosphere made it easier for me to network (another of my underdeveloped skills). I reconnected with Jay Humphreys, an associate at Mt. Wilson Ventures and recent Anderson MBA whom I had first met at the September 2, 1999 celebration of the 30th anniversary of the Internet, held at the Anderson School. Jay said to contact him when I got back to Los Angeles. Associates in venture capital firms are very important allies in the funding process. While not the decision makers, they are the ones who help clarify the personalities of the principals and the pitfalls and sweet spots of each venture firm. If an associate rejects your idea, you probably have saved yourself a lot of time. If an associate wants you to meet with the venture partners, it is very much in the associate's interest to have you as well prepared as possible. This alignment of interests can be enormously valuable.

While networking may not be high on my list of abilities, the friends and colleagues accumulated over the years are a great resource for any academic thinking of venturing out.[20] I spent Wednesday evenings that late summer of 1999 sailing with Bryce Benton. Bryce, a friend, seasoned entrepreneur, and angel investor, was the CEO of his own investment vehicle, and also the COO of another small international company with industrial-sewing and assembly plants in Juarez and Mexicali, and small-machine production and high-end label printing shops in Southern California. I served on the advisory board he formed to help ensure that his marketing and financial plans

---

[20] See Shane, Scott, and Toby Stuart (2002), "Organizational Endowments and Performance of University Start-ups," *Management Science*, 48, 1 (January), 154-170, for an empirical account of the important role that such human capital plays in the success of university-based ventures.

could withstand outside scrutiny. In turn, he very willingly shared his perspectives on the new-venture initiation and any specific issues – as long as it didn't interfere with sailing maneuvers such as *tacking* or *coming about*.

As the time to contact Jay approached, Bryce reacted insightfully to the draft business plan as it emerged from my work with Van and Teri. The day before my lunch with Jay, I practiced a pitch to Bryce, with Van listening in on the critique. There are four kinds of risks a business plan must confront, according to John Doerr of Kleiner Perkins Caufield & Byers:

a. Technology Risk – will the technology do what is promised?
b. Market Risk – is the potential market big enough to matter?
c. Financial Risk – can the required capital be raised?
d. Human Risk – can the venture gather the human capital to master both the technology and the management needed to bring it to market?

For that September 30 lunch with Jay, we had barely a sketch. Our name, Strategic Decision Corp., helped focus attention on the strategic use of customer data. Our motto, *From Analysis to Action*, underscored our desire to be the antithesis of "analysis paralysis." We constructed a value proposition:
• translate the billions of dollars businesses have invested in customer databases into bottom-line revenues;
• create an integrated team of strategy, marketing, and computer-technology experts; and
• develop state-of-the-art datamining systems – rule-based data miners that are ready for enterprise-level implementation.

The first bullet point of the value proposition targeted the market risk. The second bullet point targeted human risk, while the third targeted technology risk. Capital risk, at this point, was a complete question mark.

We could point to a top management team. I was chairman of the board – obviously responsible for overall management as well as the models that drove our approach to technology-enabled marketing. Giovanni Giuffrida, CTO, was the master of datamining and software development. As an advanced doctoral student in computer science, his knowledge was much broader than we had tapped in our prior engagements. David VanArsdale, vice president of operations, brought experience in and an understanding of general management, as well as the specifics of network-based systems, such as the one he helped design and implement in the then-new Anderson complex at UCLA. Teri Connelly, vice president of business development and marketing, brought experience in using customer databases for marketing in telecommunications and utility settings, and expected to be able to deliver clients in those areas.

I had learned of the importance of boards of directors in my days as director of the UCLA Management in the Arts Program. Harold Williams, then our dean, later head of President Carter's SEC and still later head of the Getty Foundation that built the Getty Museum in Brentwood, tried to teach me the fundamentals of working with boards, as well as the role of boards of directors as boundary agents for a company. So for our board of directors, I already had designated Steve Mayer, Penny Baron, and Bryce Benton – with me as chairman. Steve was one of the first eight at Atari, part of a group he referred to, self-deprecatingly, as Nolan Bushnell and the seven dwarfs. As the inventor of the first programmable home-video-game console (the Atari 2600) as well as the Atari 400 and 800 computers, Steve learned what hypergrowth really meant for technology companies. He left Atari to head up Warner Labs in New York, and watched from afar as Atari imploded. The same day Warner paid Jack Trameil to take Atari off its books, Steve convinced the company to seed him the funds to jettison Warner Labs into an independent company that became Digital F/X. Steve's name and reputation got him in to see 150 venture capitalists before finally getting funding from Kleiner Perkins Caufield & Byers. Digital F/X was Vinod Khosla's first venture-capital deal after leaving Sun. I knew Steve's experience would be extraordinarily valuable to me. Penny Baron has a PhD in social psychology and spent years on the University of Iowa Marketing Faculty before taking a leave-of-absence to be one of the three founders of Information Resources Inc. (IRI), one of the world's largest marketing-research firms. Penny deeply understood how transaction data from modern retail environments (i.e., scanner data) had to be processed and presented in order to support decision-making by retailers and manufacturers. Penny left IRI to help design the next generation of scanner information systems at AC Nielsen, and when it chose not to build them, she co-founded *efficient market services, inc.* (*ems, inc.*) to do it. Having built two large companies based on customer data and marketing models, Penny had decades of relevant experience for me to tap. We had worked together on occasional projects since the early 1980s and had worked closely on the development of PromoCast – the forecasting project that had brought Giovanni and me together. With these individuals and Bryce Benton, whom I've already mentioned, I felt I had the nucleus of a board that could advise me well and bridge to a number of the constituencies important to SDC's future.

I also recruited an initial panel of academic partners:
- Prof. Donald G. Morrison, Leonhard Chair, Anderson School at UCLA
- Prof. Dominique Hanssens, Knapp Chair, Anderson School at UCLA
- Prof. David Midgley, professor of marketing at INSEAD
- Prof. Bart Bronnenberg, assistant professor of marketing, Anderson School at UCLA
- Prof. Akihiro Inoue, associate professor of marketing, Kwansei Gakuin University

These friends and colleagues possessed deep theoretical, methodological, and practical knowledge about marketing science. My connection to these and other leading academics gave our small company intellectual capital far beyond its immediate boundaries.

Our products included:
- KDS (Knowledge Discovery System): a rule-generating dataminer implemented as a superset of SQL. The *Management Science* article on KDS, already accepted at that point, provided the *imprimatur* of a highly respected, peer-reviewed journal – a valuable point of differentiation in an arena dominated by arcane heuristics only superficially described by our potential competition. We also published in the computer-science (IEEE) literature.[21] While publication put the method into the public domain, the software implementation was exceptionally clever and efficient, developed without university resources, and privately held. Thus, we didn't have to worry about any university claim to the intellectual property.

---

[21] Giuffrida, Giovanni, Lee G. Cooper, and Wesley. W. Chu. (1998), "A Scalable Bottom-Up Data Mining Algorithm for Relational Databases." In *10th International Conference on Scientific and Statistical Database Management* (SSDBM '98), Capri, Italy, July, IEEE (Institute of Electrical and Electronics Engineers) Publisher.

- Noah: a datamining algorithm optimized for cross-selling applications on customer databases. The working papers on this method were much less advanced than KDS at this point.[22] Again, the mathematics would ultimately be public but the implementation private, for the same reasons as with KDS.
- Future Products: software and ASP (application service provider) capabilities for datamining of Web sites and real-time translation of click streams to customer-support initiatives. At least by labeling this as "Future Products," we correctly marked it as vaporware.

Our basic go-to-market strategy involved delivering two tiers of services: Gold and Silver. The Gold service represented a classic "bowling-alley" strategy (Moore, 1995) in which each Gold client was treated as a custom-development opportunity, with the emphasis on creating a reference-able account. The learning from that client's particular vertical sector would aid in our developing an almost shrink-wrapped version of the application that would serve the Silver clients in that same arena. We thought of three revenue streams from each client: initial mining (software), incremental mining (ASP), and an action generator (ASP). In this setup, Silver clients would be our hypergrowth opportunity.

Communicating even this much over a lunch at the 17th Street Café was a challenge. But apparently I got enough of the message across. Jay asked for more supporting materials (the *Management Science* article exemplifies an advantage academics have in exactly this circumstance), and said he would talk to Jim Bauer, the managing partner, and try to set up a meeting with him as soon as possible. The meeting was ultimately set for October 20. The break gave me time to finish setting up my Fall Quarter classes, two sections of a course called Market Assessment – a relevant topic for anyone considering new ventures.

With the Mt. Wilson Ventures presentation approaching, the prospect of funding became serious enough that I asked Bryce to recommend an attorney with Internet experience. He offered the name of a very good attorney – in Pasadena. For those unfamiliar

---

[22] Giuffrida, Giovanni, Wesley W. Chu, and Dominique M. Hanssens (2000), "Mining Classification Rules from Datasets with Large Number of Many-Valued Attributes." *Proc. 12th Int'l Conf. on Extending Database Technologies (EDBT)*, Konstanz, Germany, March.

with Southern California traffic patterns, there is no sane path from Santa Monica to Pasadena. If, by some magic, you can get there in under an hour, you'll never get back. I decided to call another friend, Bud Pennington, for a recommendation.

Bud is the brother of one of my very best friends, Skip. Skip, three others, and I had been an inseparable gang of five in high school and had stayed close over the years. Bud and I reconnected when I first joined the faculty at UCLA just after my 25th birthday in 1969 and he, at 28 a recently resigned Air Force captain, was in his second year of UCLA Law School. Bud went from *UCLA Law Review* into the Beverly Hills entertainment law firm Kaplan-Livingston, and then jumped out to form a firm with Kenneth Fisher a year or more before Kaplan-Livingston imploded. Fisher Pennington became a major player in Hollywood – innovating 25 years ago by taking a slice of the deals the firm created, rather than an hourly fee. Although Bud lived 10 blocks away from me in Santa Monica, I referred to the distance between our homes as being 10 blocks and $20 million. Our contact was mainly in the family circle, rather than the social circle or business. His mother and mine were the ever-present parent hostesses trying to herd a gang of overzealous teens. But in January of 1999, Bud and I lunched at the UCLA Faculty Center to discuss his role and experience in creating a proprietary, copy-protected, distribution alternative to DVDs. A sketch of its creation, funding, and too-early demise appears in Chapter 5, "Smart Money." Despite knowing this history, I thought of his practice as being otherwise exclusively Hollywood. I was surprised when, in response to my phone query for a recommendation, he said, "Coop, we're all over that. We have a whole Internet practice. I'll put you in touch with the head of that practice, Paul Brendl. You'll have to go through our screening." I knew that Fisher Pennington stayed small and extremely profitable by being very careful about client selection and regularly culling its client list. So, once over the initial surprise at how little I knew about the firm's Internet efforts, I realized the screening of client candidates certainly fit its approach.

I expected a call from Paul Brendl. When it hadn't come in a week, I came as close as I could to jumping on the stage. I called Paul and told him that Bud had said to contact him. I gave him the basic idea on privacy and our way of using datamining to personalize e-commerce. He immediately got the idea and probed gently but deeply into whether or not I had the technical muscle and practical sense to carry it through. He then explained that they had a three-stage screening process, and that I had just passed the first one. Next would come a live presentation to a subcommittee of their Internet practice; if I passed that hurdle, the subcommittee would present the case in private to the whole practice. I should expect to present for 45 minutes and get 15 minutes of feedback. He called after my afternoon class on October 13 to tell me he'd set up a meeting the next morning with the subcommittee.

I met with Brendl, a partner and head of the Internet practice; Kathleen Hall, a partner; and Leeann Swit, who developed and managed the IT infrastructure for the firm. We congregated in the main conference room at Fisher Pennington – a room designed to impress people far more important than I. The oval, marble conference table is probably 25 feet long, with a long sideboard for serving on the north wall and a wall of sofas on the south, above which hang an absolutely stunning series of Miró gouaches. I positioned myself facing north so I wouldn't be tempted to spend my time staring at the irresistible Mirós. I spoke from the simple black-and-white PowerPoint presentation. I covered the value proposition, goals, market size and major trends, how the company planned to attack the market, the top management team, our academic partners, current products, product strategy, what we wanted in financing, and our marketing communications message. I augmented this with five plots showing projected 48-month cash flow, income, total personnel, payroll costs, and non-payroll costs. The augmentation since the lunch with Jay Humphreys mostly focused on adding specificity to the market-size component – citing IDC reports on the growth of *Knowledge Management* in general, with the software and services part, into which we fit, expected to grow to $9.7 billion by 2003 – and discussing how our offering aligned with the major trends IDC had identified. Calculating lifetime value of customers was very much a part of the shift in marketing-science practice from considering brand equity to thinking in terms of customer equity. The projected growth in small online business fit right into our hypergrowth projections. The notion that personalization/customization was the "ante" for successful e-commerce sites positioned SDC to be a candidate for the foundation of future e-commerce – a required part of the infrastructure, the *tornado* as described by Geoffrey Moore's (1995) writing.

The four of us huddled around the east end of the giant table, projecting the presentation onto a portable screen rather than the monster drop-down screen at the west end. The level of interaction pleased me, so different from my statistics lectures that were often greeted by total silence and blank stares. Despite the dialogue over issues as they arose, I still managed to get through my slides in the allotted 45 minutes. These are busy people, but the expected 15-minute feedback kept going – partly probing to find areas I hadn't considered. Here, doing your homework is important. First, all the financial projections for the company were based on spreadsheets with expected client-by-client support needs, costs, and expected revenues. I could, if needed, drill down to the base assumptions behind each summary – showing how each final number was produced. I've seen too many business plans that position a new venture in a $10 billion market and then assert, "If we can even capture 1% of that market …" Your plan has to say how you intend to try to get the 1%. Second, I had brought (but not shown) a supplemental presentation sketching out Strategic Decision Corp. in the 11 dimensions Slywotzky[23] uses to characterize companies and the policy decisions they face: Fundamental Assumptions, Customer Selection, Scope, Differentiation, Value Recapture, Purchasing System, Manufacturing/Operating System, Capital Intensity, R&D/Product Development System, Organizational Configuration, and Go-to-Market Mechanism.

As the feedback went on, I felt as if they were trying to align their expertise to the needs of SDC – always a good sign. Add color – it's too much like a stodgy academic talk. Spell out the competitive landscape better – don't pretend you are alone. Understand it is a dual sell to customers: the IT guys and the senior executives. Recognize that every proposal VCs in this area are seeing has datamining as a part. Differentiate yourself with your 30-year credentials in this area. Position your datamining as required infrastructure, as the Cisco of customer databases. Emphasize that you are only six months pre-revenue. Put in potential valuation for company in plot or chart (and make sure that valuation is big). Ask for more money – $5 million is a much more reasonable request for the needs as articulated. Realize that the VCs at that time wanted a

---

[23] Slywotzky, Adrian J. (1996), *Value Migration: How to Think Several Moves Ahead of the Competition.* Boston: Harvard Business School Press. This approach to sketching a business is discussed in Chapter 6.

10× return on this in 2-3 years. And do this in no more than eight slides.

Fisher Pennington is not a long-form law firm. By that I mean this firm is not the one that produces the detailed contracts. The firm thinks of itself as a collection of strategic advisors – putting together the pieces and establishing the alliances needed to make the entertainment business go. So many of the comments centered on who could help make SDC a success. While Paul made it clear that the entire practice would have to be presented the case and decide, I felt like I had successfully completed the second hurdle. The decision was due Tuesday, October 19 -- one day before my meeting with Mt. Wilson Ventures.

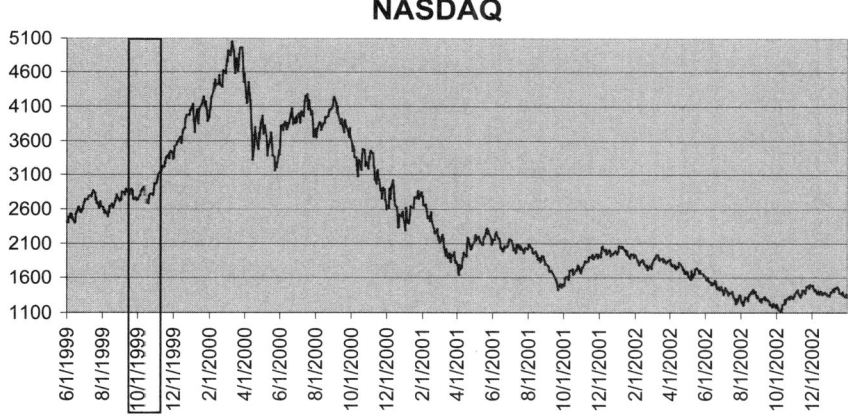

For the Mt. Wilson Ventures meeting I added color and transition effects. I moved the marketing communications message before the value proposition, added the section on the competitive landscape that listed 40 vendors of datamining software, and inserted financial information that included the funding we were seeking, the amount of time we expected to be pre-revenue, and our revenue projections for years 2 – 4 along with expenses, personnel growth, and our estimate of valuation – showing SDC would be a $2 billion company by the end of year 4. I managed to reduce the presentation to 15 foils including (two) plots. The rest of the suggestions I hoped to work into my comments.

As promised, Bud called in the late morning of October 19. I was delighted when he welcomed me to the practice, saying every one of the Internet companies Fisher Pennington backed had been a success, and money had never been a problem. Whether that was true or not didn't matter: Bud's enthusiasm was contagious. I knew that the firm's backing would make a huge difference in our prospects. We established that Paul Brendl would be our attorney, and that Paul would set up an initial meeting soon for all the involved parties. We agreed that I should present to Mt. Wilson Ventures the next day, get as much input and feedback as possible, but make no commitment. Mt. Wilson Ventures focused on early-stage investments in information technology. Its perspectives would be important regardless of the company's ultimate interest in investing.

I went into the Mt. Wilson Ventures meeting probably more relaxed than I would otherwise have been. Jim Bauer, George Sato, and Jay Humphreys attended. As I expected, Jay was silent. Jim and George were very cordial, but all business. They questioned the strong consultancy flavor of the organization. Consultancies were rarely good candidates for venture funding in their view (despite the service needs required to provide a whole product solution to early customers). They also indicated that from their point of view, no datamining firm had been a success in the last 20 years. Why did we expect to be any different? I spoke a little about real-time, technology-enabled marketing, without trying to appear defensive.[24] Mostly, I listened. The design did sound more like a consultancy than made me comfortable. Teri's idea, aligned with her experience hiring consultants in telecom settings, was that we could get a series of clients that we billed $300K per month, grow the consultancy to $200 million valuation and sell – echoing the Garage.com slogan of "Start up. Kick butt. Cash out." That wasn't quite what I had built into the financial spreadsheets, but the consultancy theme clearly resonated in the marketing communications message. As we moved closer to being a pure Internet play, this orientation fit less and less. Mt. Wilson Ventures wanted to keep in touch as our plans matured. That was a nice way of saying, "No."

The opening meeting with Fisher Pennington was set for 2:30 p.m. on October 28, at the firm's Century City offices. I scheduled a dinner meeting the night before at my house with Van, Teri, and Giovanni. At that dinner, Van coined the phrase we later used to

---

[24] They also told me to drop out the valuation information that I had just added. *Valuation* was their expertise, not mine. No harm was done, but I learned a lesson about the diversity of styles. What one party feels you must mention, another party may feel quite differently about.

brand our personalization solution. Though we didn't immediately put it into the discussion list for the next day, we moved quickly to reserve the appropriate URL, just as my wife Ann had reserved our corporate name with the State of California the day before. While we celebrated at the dinner that the fantasy of summer was becoming a reality of fall, that very reality put an ominous undertone to the discussion. Giovanni was concerned about his immigration status. With a limit to the practicum years he could accumulate, if he left HRL, he would need an H1B visa in the near future. It seemed to me that Teri was concerned (correctly) that marketing and business development would lose power in the shift from a consulting partnership to an Internet firm with Fisher Pennington helping to bring in clients. I wondered if I had led the others to believe that this would be a simple four-way partnership.

Only Van seemed to have his ego in check enough to wait and see. Coping with the severe medical problems of his recently born daughter occupied much of his time that fall. While he privately expressed gratitude that he was still a part of the endeavor, I knew every organization needs a designated adult, and that he was ours. Geoffrey Moore later wrote extensively[25] on how the evolution of economic webs enabled digital companies to focus on their core competencies and outsource other functions to specializing firms. While that was the ideal arrangement, to some residual extent all firms have administrative concerns. Van brought core knowledge in network arenas, but would also be primarily responsible for dealing with the context, rather than core, to the extent we couldn't outsource.

In the meeting the next afternoon we introduced the other three team members to Paul Brendl and Kathleen Hall, and in turn they introduced us to Edward Lennon, Josh Huffman, and Adam Kahn of Lennon & Ortega (whom we had met only over emails), along with two attorneys from patent practice at Foley, Hoag, and Elliot – a Boston firm that had a great deal of experience working with MIT faculty. As explained to me privately, Edward Lennon was expert at shaping the materials for the road show expected for our Series-A funding, and orchestrating our anticipated tour of East Coast funding sources. Josh Huffman would help craft the documents creating our

---

[25] Moore, Geoffrey (2000), *Living on the Fault Line: Managing for Shareholder Value in the Age of the Internet*. New York: Harper Business.

company and talk me through the important organization decisions inherent in that process.

From the patent attorneys we received not only a lesson in the importance of patents to venture funds at that time, but also clear examples of the value of patents covering very general areas. A patent on the design of a chair would be very general, indeed (and, of course, very subject to being overturned by prior art). The more adjectives that specialized the *chair* design being patented, the less competitive protection the patent offered (but the more likely that prior art would not exist). So a patent on a *wooden chair, with four legs cross-braced and dowel-pegged to a seating platform* would obviously be of less value to a potential funding source. I already had some awareness in this arena, having admired the patent on Steve Mayer's home-office wall for moving a pixel across a video screen in response to a joystick. Not too many adjectives in that patent. Atari had etched the patent circuitry and summary on a plaque for Steve after it contributed to $600 million in sales. We worked with Giovanni and the patent attorneys on the language that most generally described our efforts – settling on the phrase "remote controlled, secure, residential, real-time datamining agents" to characterize our approach to personalization.

As the meeting broke up, Paul called me into the smallest conference room at the end of the hall. Here, Paul and Edward sketched out to me the structure of our deal. Fisher Pennington normally took a 5% interest in its client (that wasn't diluted until the client had received $5 million in funding), plus a minimal fixed monthly fee for the strategic services, connections, and deals it set up. Because Paul thought a lot of organizational work was going to be needed, Fisher Pennington would split its normal interest with Lennon & Ortega. My understanding was that for its half, Lennon & Ortega would do the long-form work to create the company, develop a financial structure, draft employee contracts, etc. Once the company was structured and financed, we would receive a 25% discount on subsequent long-form work. I also understood that Lennon & Ortega would orchestrate the road show for the funding.[26] The enthusiasm for our efforts was very encouraging. Paul kept emphasizing *time to*

---

[26] This was the first time, but not the last time, that my understanding of a meeting's outcome was different than Edward Lennon's. I'm not implying blame – just admitting we had difficulty understanding each other.

market and how important it was to kick-start this company as soon as possible.

In phone conversations, Josh and Adam tried to convince me of the benefits of forming an LLC (Limited Liability Company) rather than a C corporation. Bryce and I had discussed LLCs during one of our sailing afternoons. An LLC had the liability protection of a corporation, but was taxed as a partnership – avoiding the dreaded double taxation of corporate profits, and meaning that the early losses expected in a startup could be written off against personal income. The downside was in drafting an arcane management agreement to parcel out ownership and control of the enterprise. I understood C corporations – having run Cooper Research Inc. (a California corporation) as a consulting vehicle for almost 15 years. Every company discussed in the *Bootcamp for Startups* was a C corp., but I could see the tax advantages, and was reluctant to go against the attorney's advice so early in our relationship. So on November 4, 1999, we became the Strategic Decision Group, LLC (a Delaware LLC) – the same day I flew up to San Francisco to present my datamining work to the trustees of the Marketing Science Institute.

The next day, I left for Baltimore to say hello to my son at Johns Hopkins University before presenting my research at the INFORMS (Institute for Operations Research and Management Science) Conference in Philadelphia. On the flight back to Los Angeles, I finally had time to reflect on the enterprise I was undertaking. Giovanni was bright and talented, but kind of a nudnik who kept pointing out what datamining could do to augment my more formal statistical models used in PromoCast forecasting – until I finally took notice. In that context I had the added discipline of working off residuals from a good forecast, with a well-defined dependent measure (the error in the number of cases of merchandise in the forecast versus those cases actually sold). I was now considering deploying these dataminers to discover patterns in massive customer databases from e-commerce sites, betting a lot of other people's money on the outcome. What was my dependent measure? Without a generalizable dependent measure, I was in essence relying on unsupervised discovery that could be flawed at worst, or at best dependent on client-by-client interpretation. I hated to do custom interpretation – it relied on *content* knowledge, when what I possessed was *method* knowledge. It wasn't scalable, as I knew our efforts would have to be to have the impact that everyone was expecting. This would not do.

I might have gone into existential shock had the answer not come to me immediately: Segment, Target, and Position – the basic modern-marketing mantra. The segmentation scheme was the dependent grid on which we could find out how popular each Web site was with different types of customers and, more importantly, learn customer preferences – datamining to discover simple rules in each segment – before recommending to customers the most popular items in their particular segment that they had not already purchased. We could call the approach SCOPE, for Segment and Customer-Oriented Preference Engine. In the e-commerce case, since we would be dealing with a site's own customers, we could use the traditional direct-marketing variables, ZAG (ZIP code, age, and gender), to segment. Commercial segmentation systems that might work already existed, such as Claritas's PRIZM®. I wasn't sure that the internal details of PRIZM® were quite what I wanted, but I was relatively sure that costs would grow to about $5 million per year to license it on the scale we envisioned. Besides, I reasoned, I had known for many years how I would design a census-based segmentation, with ZIP code as the linking key. It was the kind of undertaking that was too advanced to teach to MBAs and too practical to teach to PhD students in marketing, but fit extremely well with my academic background in psychological measurement and multivariate analysis. So I decided to create ZipSegments as an e-commerce-actionable segmentation scheme. Since it would be done from the ZIP code files of the 1990 U.S. Census, it would be actionable for all direct marketing channels, not just e-commerce. I now had the basis for the business I'd been pitching for the last month.

3.1   Building a Team

Despite my increasing belief that this company would be about technology-enabled marketing rather than having the consulting flavor of our early discussions, I knew I would need a highly talented team to introduce our application to early clients. We needed first to have referenceable accounts. Second, we needed to learn what was the minimum customization need, and conversely, what was the maximum part of our application that could essentially be shrink-wrapped for other clients. Having known numerous very talented

MBAs who have passed through the Anderson School, three names stood out: Jason, Kate, and Troy.

I first met Jason Kapp as the husband of Sara Kapp. The prospect of working on *Project Action* attracted Sara to the UCLA doctoral program in marketing. Jason decided to get an MBA while Sara worked on her doctorate. He came with a strong IT and Internet background, performed at the top of his class, and took a job with AT Kearney, in its IT Strategy practice, when he graduated in 1998. When I described the concepts behind SCOPE, he recognized it immediately as key to the Internet version of the Efficient Consumer Response initiative. The basic learning mechanisms in SCOPE provide the detailed demand indicators at the e-commerce retail level to drive decision making throughout the supply chain. His group at AT Kearney had just finished an exercise concerned with the IT role in the e-commerce supply chain, and concluded that all the pieces were in place except this one. I had hit the sweet spot. We arranged to meet mid-afternoon on November 19 to discuss this further.

Kate Garrett actually graduated from the Arts Management Program at UCLA earlier in the 1990s – long after I had stopped directing it. I knew she had excellent online marketing experience through her consulting work in which she brought a major bank into the Internet age, as well as earlier efforts developing a Web portal for firms seeking to understand how to do business in Japan. I was current on her efforts basically because she was married to one of my close colleagues. I pitched the idea to her in the quieter, early part of the UCLA-Washington football game that we ended up winning 23-20 in overtime. We arranged to meet for lunch on the 19th.

The third leg of this stool was Troy Noble. Troy had worked developing database decision-support products for *ems, inc.,* Penny Baron's company for which I had done PromoCast. When he left to go to business school at UCLA, mutual friends at *ems, inc.* insisted that he drop by my office to say hello. At the beginning of the Fall Quarter 1997, he stopped by for a brief but interesting chat. Moments after he left I told myself I had to get him involved in my work on Project Action. With decent econometric skills from his undergraduate studies at Michigan and his practical experience in decision support for information-rich environments, I knew I could use his talents. Sara Kapp had left Project Action after a year, when her research interests veered more toward behavioral decision theory than high-technology marketing, and eventually transferred to the UCLA Psychology Department to work on a PhD in cognitive psych. The Project Action staff was basically down to me and a brilliant UCLA undergrad, applied-math major, Laura Baron. Yes, Laura is Penny's daughter – just to keep this all in the family. Laura had helped identify and tutor me on Bayesian networks as the third and ultimately successful candidate for use with the strategic-planning framework that was evolving out of Project Action. Troy had the quantitative skills to understand how I was using Bayesian networks, along with the practical experience in product development. His upcoming MBA core courses would supply the rest of the needed ingredients. My experience with Troy as my research assistant convinced me I should try to recruit him to the Strategic Decision Group team. He had graduated only five months earlier. This was not a good time to try to pull him out of his job as a Gartner Group consultant, but I was thinking more of our needs than his career. If he thought it was bad for his career, he could always say no.

I also contacted Pam Pennington, who had an independent human resources consulting practice. Yes, Pam is Skip's daughter and Bud's niece – just to keep this all in the family. She had helped L90 through explosive growth, and was happy to supply prototype personnel manuals, offer letters, employee non-disclosure agreements, and general advice on how not to screw up by promising future employees things you cannot deliver. Her can-do attitude and cheerful persona were complimented by a deep understanding of how critical early hires are for a startup. The founders may be most determinant of the corporate culture of a startup, but the first wave of employees largely determines if the entrepreneurial spirit can be sustained in early growth or a bureaucratic mind-set will take over. So I went into my meetings with Jason and Kate intent on communicating the idea and my enthusiasm for it, but tempered by a need to hear their practical concerns. If you tacitly communicate the salesman's sense that you will not take no for an answer, you run the risk of disenfranchising the articulation of real roadblocks to success. Just as marketing talks about *hearing the voice of the customer*, the marketing view of the firm says you must hear the voice of the labor markets.

Meanwhile, Fisher Pennington was busy trying to get seed funding to kick-start the company. Bud had suggested to Fred Hart that the two of them go in together. Fred was a long-time client and friend of

Bud's. His $1.2 billion net worth put him several places ahead of Steve Jobs in the 1999 Forbes 400. Forbes.com summarized Fred in 1999 as the Hart department store scion turned producer, with 50 major motion pictures to his credit, his current hit was the fastest romantic comedy to cross the $100 million mark at the box office. This success was a good distraction from ongoing tumult at his rap music label. Fred moved to Hollywood after clash with half-brother forced liquidation of family's retail, real estate and media fortune. Fred was known as a big Democratic supporter who stayed away from summer Hamptons scene when Bill and Hillary Clinton visited.

Regardless of whether Fred could shake the initial investment in our company out of his sofa, he did not grow his inherited fortune into great wealth by making investments simply on the word of a friend. He contacted Martin Ross to conduct due diligence on us. Martin had a 1984 PhD in information and computer sciences from UC Irvine and had long been involved in the evolving Internet infrastructure. He was area director for network management of the Internet Engineering Task Force, one of the dozen individuals who oversaw the Internet's standardization process. He wrote the first POP protocol for email and was well known to others (not to me) for his implementations of other major communications protocols. A large and impressively bright individual, he barely fit into the conference table chairs at my UCLA office when he, Giovanni, and I sat down that Wednesday morning, November 24.

The Mt. Wilson Ventures experience had prepared me for his polite skepticism over a business designed around datamining. As Paul Brendl had pointed out, every business plan then being written had huge revenues deriving from datamining the information the proposed enterprise expected to gather. Discussion of the *Management Science* article and our planned implementation convinced him that we knew a lot more about the practical use of datamining than any other startup he knew of. We had the technical muscle, but did we have a business design that could monetize it? I pulled out the 11-dimension Slywotzky description mentioned above to summarize the practical side of our vision. Doing your homework pays off. I felt that we had won a supporter by the meeting's end at 10 a.m. Martin indicated that he would go and put together his notes and have perhaps some further questions to me by Monday – after the long Thanksgiving weekend. I didn't have to wait that long. By 11:30 a.m., Paul Brendl had called to tell me I'd done something right. Fred seemed willing to co-invest with Bud. He wanted to meet me the following week.

I am not at all opposed to great wealth. Other things being equal, it's a wonderful outcome. But I do not see that as the goal of entrepreneurship. Many successful entrepreneurs see business as a game to be won, with wealth as the scorecard. I do understand and value that point of view. But I see the entrepreneurial spirit as the business expression of a creative impulse. In business school one sometimes gets the impression that the richest person is the smartest. I don't agree. I'm a lot more comfortable with Guilford's theory that articulates 150 components of intelligence.[27] I doubt that something as rare as great wealth can reliably be related to any of these. Many feel that the wealthiest are the happiest. I think Dickens was closer to the truth when he wrote for his hapless character Wilkins Micawber in David Copperfield:[28]

> 'My other piece of advice, Copperfield,' said Mr. Micawber, 'you know. Annual income twenty pounds, annual expenditure nineteen nineteen and six, result happiness. Annual income twenty pounds, annual expenditure twenty pounds ought and six, result misery. The blossom is blighted, the leaf is withered, the god of day goes down upon the dreary scene, and - and in short you are forever floored. As I am!'

I sped through many such thoughts as I tried to rid myself to the myths, prejudices, and assumptions I held about billionaires before I blew a big opportunity.

To prepare for the meeting Giovanni, Van, Teri, Jason, and I had a dinner meeting at Teri's – our first group meeting with Jason in attendance. I felt that Teri attempted to control the agenda. She wanted to do great things for the company, but as the company steered into an arena much more remote from her experience, that seemed less likely. I had tried to align Teri more with the emerging Internet focus by urging her to attend the first Personalization Summit in San Francisco earlier that November. Everything was voluntary at this point, and when she found reasons not to go, I did not complain. I never mentioned it again. Despite the awkwardness we felt as we stumbled through our agenda, I felt prepared by the end

---

[27] Guilford, J.P. (1967), *The Nature of Human Intelligence*. New York: McGraw-Hill.
[28] Dickens, Charles (1850/1948), *David Copperfield*. Garden City, N.Y.: Literary Guild of America.

to handle the uncertainty of tomorrow's meeting. In phenomenological inquiry they talk about sifting through personal foibles and theoretical predilections before entering authentic dyadic communication.[29] The idea is you get all that stuff out and *bracket* it by then saying to hell with it. You then try to stay in the moment – listening and reacting without bias. Perhaps a better analogy is the UCLA basketball team, after stumbling through practices and the anxious highs and horrid lows of the conference season, finally deciding to go out and play Stanford as if it's a pick-up game – and beating them on their home court.

Fred's office was a few floors below the Regency Club in a Westwood Village high-rise. I expected Fred to be there and maybe Bud, but was a little surprised when I was introduced to Jay Hillis (Fred's CFO) and Len Steiner. Bud once mentioned Len Steiner's name as someone whose talents and connections might be very helpful to this new company. Len had been the president and CEO of First Virtual Holdings, Inc., the first financial and marketing company to create an authentication system that enabled safe global electronic commerce on the Internet. Martin Ross, one of the other founders of First Virtual, had obviously helped to spread the good word.

The office was as impressive as you probably imagine. After introductions we settled into the seating area in the south 40. Fred took the chair on the west at the head of a coffee table, Jay on his left, and Len on the opposite end. Bud took the sofa on the south, while I sat between Jay and Len. I barely had to open the discussion when ideas began to fly. It felt like a verbal analog of the old ping-pong ball contagion experiment. You had a room full of mousetraps, each with a ping-pong ball for bait. I tossed out the first ball. Every place it bounced it set off another ball, with each of those setting off others as they bounced. By the time the last ball bounced we had firmly aligned the company's future with real-time, technology-enabled marketing optimization, all four were excited to invest, and all four wanted seats on the board of directors.

---

[29] See, for example, Massarik, Fred (1985), "Human experience, phenomenology, and the process of deep sharing," in Tannenbaum, Robert, Newton Margulies, Fred Massarik and Associates (eds.) *Human Systems Development*, New York: Jossey-Bass, 26-41.

While stoking their enthusiasm for investing I was careful not to commit to the four board seats. Four already-proposed directors, plus four for the A-Series investors, created too big a board to start with and gave too much power to the investors. By the B-Series, investors would have a majority of board seats and might cite these four seats as precedent for the number of seats that investors would demand. This was very much on my mind when Bud and I set a meeting for Saturday at his house.

At the Saturday meeting with Bud I confessed my discomfort with giving four seats to the Series A. Bud said, look Lee, this is your company. We need you to be comfortable with the board. I said that two seats were okay for this round with the understanding that future rounds would get only one each. Bud proposed that Fred and Len Steiner take the seats. I told him how important it was that he took one. Bud was the guy who was orchestrating all the money and connections. Bud said that Len Steiner would be better for the company at this stage than Fred Hart. We agreed on that board of six. It was an even number, but as Penny Baron pointed out later on the phone (and on numerous subsequent occasions), a board that can't reach consensus at this point is doomed, so even-versus-odd numbers did not matter now. How much money would be in this round grew over the next two weeks from $875,000 to $1.25 million, as we provided more concrete numbers for our burn rate, Bud found the positive reaction to the concept among his partners and associates, and we strategized about how long we should plan to last before going for more money.

Everyone wanted us to keep going full steam while the deal details were developing. Bud said he would cut a check for $50,000 and Fred would match, to enable us to open a bank account, start looking for office space, buy just the beginning equipment, and make the commitments to the key employees. When I stopped by Bud's office the next week to pick up the check I was surprised there were no papers to sign. I asked Bud if he wanted a receipt. He started to explain his "no" when Harrison Ford called him from the set. While Bud handled whatever occasioned the call, I wrote out a note on plain paper saying simply "Bud gave me $50,000" and signed and dated it. Fred wired his match directly to the bank the next morning, as soon as the account was opened. I was entering a different world.

Meanwhile, full steam ahead meant just that. Backing up to the Sunday after my Saturday meeting at Bud's, we scheduled an all-day planning session including the existing team plus Jason, Kate, and Troy. Everyone was still working their day jobs, but was committed to this venture.

Before that Sunday meeting I had to deal with Teri. The only down point for me during the fabulous interchange with Fred, Bud, Len, and Joe was after one fascinating suggestion by Len my private thought was how could I position this to Teri. Worrying about Teri's feelings during a meeting that was so positive for our prospects was not the right thing to do. I shut her out of my mind for the rest of the meeting, and knew Teri was not part of the future of what was evolving. It was not the same as firing someone, since no employment contracts had been offered yet, but Teri had become a friend over the course of a year. That made it more difficult. It wasn't about lack of skills, but about the alignment of skills and experience with the needs of the company. Communicating this was one of the hardest tasks I'd faced so far. But face it I did, in a private meeting just before the scheduled Sunday planning session. I had to accept my share of the blame for not judging correctly the needs of the company from the beginning. Being upfront about my culpability for misjudging the needed skill, while being clear and final about my decision, left little or no room for Teri to counter argue. She expressed some of her disappointment, but didn't try to reposition or spin the situation. I was sure she had other alternatives, but we did not speak about them. She accepted the news calmly without protest. If you can't face such hurtful situations without losing your own humanity, don't start a company.

I began the planning meeting by telling them what had just transpired with Teri. I hoped that Kate would take over the marketing function, but not business development, since in many ways Fisher Pennington changed how that would be handled. Our task was to put real substance behind the general ideas that brought us together. Gone are the days of chalk and blackboard, so I tossed the black marker to Jason, sent him to the whiteboard, and let him start this next planning phase. I am not good at leading a discussion and recording that discussion at the same time. I tend to turn it into a lecture, and that is not what I wanted here. With Jason, Kate, Van, Troy, and Giovanni, a great deal of management talent existed in that conference room. Letting them contribute would make a better business plan than I could ever create. Being a less central participant let me reflect on the quality of the effort, the appropriateness of the designs, as well as what was not being discussed. From the side I can sense if the wielder of the marker is working a common agenda or a hidden one. I lived through the process experiments of the 1970s – T-groups, sensitivity training, leadership training, open-systems planning, and other experiential approaches to organizational design and development. That experience made it easier to tell when personal needs are being put ahead of company goals. Jason was there for the common goals and so was everyone else in that room. Personal needs are always present, but I sensed an alignment of those needs and goals with our common agenda. I had all the ingredients I needed to plan our future.

We started with the vision: From Analysis to Action using remote controlled, secure, residential, real-time datamining agents – characterizing our approach to personalization. Our mission was to create the infrastructure for the strategic use of customer data.

General roles for the first 60 days came next. Kate would work on corporate identity, the look and feel of the demo, planning our Web site, the content of our marketing materials, and public relations. Jason would work on the burn-rate model. Troy would develop the database needed for the demo. Van would find space, plan out our IT infrastructure, and develop corporate accounting. Giovanni would develop the demo engine. I would continue work on ZipSegments as well as the overall organizational structure, funding, and board structure.

Then we turned to the issues in marketing, product development, operations, and finance. We had to make pricing decisions: licensing fees, support-service fees, and participation structures (i.e., value-based pricing in which we take a slice of the value we create). We also needed to decide who was our ideal client, what we wanted in the executive demo and the technical demo, and to explore closely who were our competitors. We deferred, until later, consideration of other revenue streams (e.g., co-branding relations and possible royalty streams from the intellectual property). Regarding the product itself, we discussed front-end issues such as a dashboard to report thumbnail views of system status, a reporting suite to drive off-line management decision making, and the form of the Web interface. For the back end, we considered what was in the starter package.

Certainly SCOPE and ZipSegments would be there, but we also needed to incorporate purchase-event feedback into our recommendation logic. What part of our existing datamining software would actually end up in this recommendation engine needed to be determined, as well as related issues in data integration. The structure of the services we would offer had to be spelled out.

Discussing who was our ideal first client was very instructive. That client should have an existing customer database, own its customer in the sense of having rich enough data to enable it to serve that customer better, be a dotcom company involved in exchange of funds, already be in operation, have one product category that emphasizes multiple purchases of consumer durables by repeat customers in that category, be from one store, and be big. CD music stores fit that description, as well as online toy stores, kid stores, and many other e-retailers.

We laid out an ambitious time-line for our efforts. It was December 5, 1999 and we expected to have seed money by December 17, be in full-time office space on January 3, begin our alpha testing by January 15, have a publicly viewable demo by February 1, sign our first client and begin our Series-B road show by March 1, and be ready to IPO by January of 2002. Such were the optimistic times in which this venture started.

Edward Lennon joined us in the early afternoon. He tried to orchestrate anew something vaguely related to what we had already finished. In this session and others that followed, I began to realize that a chunk of the service that his firm was meant to provide was being provided by the MBA talent now assembled in the company. I assume that Paul Brendl saw in me an academic with a great idea and no clue about practical business. Edward and his associates were supposed to add the business savvy. But I wasn't alone anymore. While I admittedly did not have the go-to-market sales force – be it in the form of a CEO or top business-development person – the management talent I assembled was far more capable than Paul or Edward had imagined. I still needed the long-form work that I thought Lennon & Ortega was to provide for its equity share. So we all tried to indulge Edward's desire to orchestrate, while listening in case uncovered areas were revealed.

With the end of Fall Quarter classes approaching I had a break until January 10. I finished selecting the variables from the U.S. Census that would be the basis of ZipSegments, secured the help of Akihiro Inoue (my former doctoral student who was visiting UCLA at that time) to look at alternative ways to pre-process the data and do the number crunching, worked with the Boston attorneys on patents and the immigration attorneys on Giovanni's H1B visa, reviewed the employment contracts the Series-A investors were insisting on for key personnel (I thought all employees would be employees-at-will), and started considering the go-to-market team.

Speed to market was a major concern. We needed either a CEO who could handle the CEO-to-CEO sales, or a high-powered business-development person, or both. Bud suggested that Len Steiner might be a CEO candidate. In the short time I'd known Len I found him to be a true gentleman and savvy business man (a rare combination), as well as very well connected on the side of the wired world that dealt with economic transactions. We meet at the W near UCLA at 6:30 a.m. on the last day of classes. He was clear that he wanted to help the company however he could, but did not feel he was the right CEO for us. Bud also arranged an end-of-the-day meeting with exactly the kind of high-powered business-development person we would need. Such people cannot wait while you build your product. By the time we had something for him to sell he would be long gone into other pursuits. Bryce and Len both brought forward possible CEO candidates with many positive skills, but not a great fit for our needs. I would remain acting CEO until we could fill this role.

Business-plan development continued with the core management team. By transforming my living room into a conference room complete with whiteboards, we became something close to a garage startup. Mostly we worked on specifying the functionality of the product and designing how the demo would capture it. Giovanni's friend, Fabrizio diMauro, showed up from New York and impressed us all with the speed of his code development. We saw evidence in line with Frederick Brooks's view that a good programmer is 10 times as productive as a poor programmer.[30] Fabrizio may not be the one you want documenting code, but as a developer, his quality and quantity are hard to match.

---

[30] Brooks, Frederick P. Jr. (1995), *The Mythical Man-Month: Essays on Software Engineering Anniversary Edition.* Reading Mass: Addison Wesley Longman, Inc.

3.2 Dealing with UCLA

With plan development and code development well under way, I turned my attention back to employment contracts. The Series-A investors wanted me and other top personnel under contract, as well as to secure patent assignments from all parties. Being a full-time faculty member, and intending to stay as such, intellectual property and conflict of commitment were concerns for me.

My prior interaction with UCLA over intellectual property concerned copyrights rather than patents. In 1988 I developed some decision-support software called CASPER (Competitive Analysis System for Promotional Effectiveness Research) as a complement to a book I co-wrote on market-share analysis.[31] I used about $6,500 in gift funds, mostly given to me by the Marketing Science Institute, to support three students who worked on the code at different times. Since CASPER won the Grand Prize in an Ashton-Tate Framework Applications Contest, co-sponsored and judged by *PC Magazine*, I thought it was appropriate to share any revenues from the software with UCLA. Being these were the early days of dealing with such issues, the Office of Contract and Grant Administration referred the matter to the Software Advisory Committee, which concluded:

> The committee has advised that your use of University gift funds identified in your proposal to support students to assist your development of CASPER did not constitute the creation of a copyrightable work subject to University ownership under UCLA Copyright Policy 960, provided that the students derived significant educational value from their participation in the project.[32]

If software developed on university equipment and university gift funds wasn't a problem for UCLA, then Giovanni's software, which was not developed on UCLA equipment and not funded with any university funds, was not a problem in my mind. When asked by Paul Brendl if I should check with UCLA on this, I said you shouldn't poke a sleeping lion with a stick and ask if this was its lunch.

---

[31] Cooper, Lee G., and Masao Nakanishi (1988), *Market Share Analysis: Evaluating Competitive Marketing Effectiveness*, Boston: Kluwer Academic Publishers.
[32] Memo dated November 28, 1988, from Wade A. Bunting, Intellectual Property Officer, UCLA Office of Contract and Grant Administration.

Conflict of commitment was another issue. The university policy on consulting was appropriate:

> The University of California encourages its faculty to participate in activities that contribute to their profession and to the outside community. In general, permissible consulting is any professional activity that is undertaken with an outside party consistent with the full performance of a faculty member's primary University obligations and that does not interfere with his or her full time commitment to the University...Faculty members and other researchers engaging in consulting activities may sign personal consulting agreements with outside entities as long as such consulting does not utilize University facilities, does not interfere with teaching and research responsibilities and time commitments to the University, and does not purport to take precedence over prior obligations to the University, including obligations associated with intellectual property arising before, during, or after the consulting period.[33]

Under the policy at the time, faculty reported annually in October on the outside professional activity of the prior year. One day a week was an acceptable level of consulting during the academic calendar. Other arrangements were possible, but largely at the discretion of the dean. I did not believe our former dean would use his discretion in my favor. Our former dean was then new to the job, but had already signaled his hostile predilections over an incident involving an Anderson Computing & Information Services (ACIS) employee who left to join Strategic Decision Corp. Carol Frittoli was an excellent manager of user relations in ACIS. For two years running she received superior job evaluations, but no bonus, while others with poorer evaluations got bonuses. So she left in disgust. But the director of ACIS accused SDC of poaching employees and told then new dean that SDC was the cause of many problems at ACIS. The former dean bought the story and accused us of poaching, and indirectly threatened legal action. First, we didn't poach. Second, UCLA employment contracts don't address poaching. So even if we did poach, the dean had no legal remedy. Regardless of the facts, I had to operate as if I had an enemy in the dean's office. That only reinforced my feeling that I had to do this within the existing policy. No special deal would be cut for me.

---

[33] http://www.ucop.edu/ott/consult.html

So I contracted to devote one day a week to this new company, adding an unnecessary, but convenient, additional layer of insulation by having that contract written between my consulting company, Cooper Research Inc., and Strategic Decision Group, LLC. Whether I could drive the company on one day a week was unclear, but I felt the closer I could get to that goal, the better for the company. Conflict of commitment is an important issue, but after I had spent 30 years of sustained research productivity, no one was going to look over my shoulder or get me to punch a time clock.

Now that I'm writing a book about the experience, I could retrospectively claim that any time over my allotted one day a week was research time. This was not on my mind at the time, and may have created more problems than it solved. I actually considered my involvement with Strategic Decision Group to be outside of my scope of work at UCLA. I was an academic, focusing on statistical-methods development, not a practitioner. As mentioned, even the development of ZipSegments, which was as close to my traditional scope of work as I got, fell in that ambiguous region between being too advanced to teach MBA students and too practical to teach PhD students. I think most management faculty feel this way about their consulting, since they know they are extremely unlikely to get any of it published, and it might diminish their academic standing if they did succeed in publishing. You don't get tenure at UCLA by publishing consulting projects.

The scope-of-work distinction matters when you consider patent assignment. The university policy here is narrowly drawn:

> All employees of the University sign a Patent Agreement/Acknowledgment as a condition of employment. Under the University Patent Policy, employees agree to disclose all inventions and patents to the University, and to assign them to the University, **except those resulting from permissible consulting activities that do not use University research facilities, and do not relate to the employee's scope of employment.**[34]

---

[34] Ibid. Emphasis added.

I was solving a different class of problems than I had ever solved before, and it was fun. Social sciences, liberal arts, and humanities faculty have here one of their few advantages over laboratory researchers, who probably don't have as much latitude. So I had no problem signing the patent agreement covering this kind of work.

The University of California regulations changed 18 months later. The implications of these changes are discussed in the final chapter.

Recently, after my patent agreement had expired, I tried to find out what UCLA thought was included in my "scope of work." The personnel director at the Anderson School was somewhat perplexed that she didn't have a ready answer for such a basic question. She asked me to write the request in an email that she would forward to the UCLA Academic Personnel Office. Two weeks later, that office came up empty and indicated that I should ask Andrew Neighbour, the associate vice chancellor for research who heads the UCLA Office of Intellectual Property Administration. Quoting from his April 3, 2002, email reply:

> Regarding "scope of employment" – I also have not yet found an acceptable definition of same. In fact, the guidelines for consulting are several orders removed from the over-arching requirement that faculty assign and disclose all inventions made during their employment at UC. This is defined in the legal contract signed by all employees. The patent policy and subsequent guidelines serve to confuse this obligation by introducing
> terms such as "permissible consulting" and "scope of employment" without providing further definitions of these terms or definition as to who determines whether the inventions fall within or without these carve-outs. In practice, for a University official to make these determinations, it is necessary for the faculty inventor to first disclose "all" inventions so that the University can determine whether it has a claim of ownership. It should be remembered that we the University and the faculty member are subject to federal law to report all inventions made using federal grant funds, and if sponsored research contracts are involved, to report new inventions to the sponsor. Many foundations have similar requirements. What is interesting is that there are no guidelines as to what basis the University will determine whether or not to assert ownership.

I emailed back, in part, that the practical necessity to disclose *all* inventions so the university can determine whether it has a claim to ownership is potentially a very explosive issue. If I tell my neighbor a better way to water his lawn, without disclosing this to the university, I might be fired for violating this interpretation of the reporting requirement. Absurd, of course, but he had already acknowledged that there are no guidelines as to what basis the university will determine whether or not to assert ownership. Thus, the university might assert ownership over my lawn-watering system. The issue becomes explosive when the faculty realizes how it imposes a prior restraint on freedom of association and freedom of speech. It becomes explosive when industry finds that all the possible inventions or patentable uses discussed in private with faculty consultants are subject to university disclosure.

He replied, "… I can't imagine any real situation where we would want to exert our rights to an improved lawn-watering system (unless you are an environmental engineer!)…" Which, of course, brought us back to the unresolved issue of *scope of employment*.

Dr. Neighbour's questions to UCOP in this area resulted in him being appointed to head a task force to think through the relevant issues.

3.3   The First "Public" Business Plan

The final Series-A negotiations are described in Chapter 5, "Smart Money." Suffice to say we ended up reforming as a Delaware C corporation essentially at the same time we closed the $1.25 million Series A on January 14. With money in hand, we moved into 2,300 square feet of office space in Santa Monica, and set to the task of completing the demo and the first public version of the business plan to present to our board of directors meeting on February 4, 2000.

Any plan has to cover John Doerr's four basic areas of risk: technology risk, market risk, financial risk, and human risk. It had better start with an executive summary that, while longer than the elevator pitch, is short enough to be read quickly, and compelling enough to get busy people to read on. Our first executive summary follows:

Executive Summary

The Strategic Decision Corporation was founded in 1999 to fulfill the demand to translate advanced customer analysis and profiling into real-time marketing actions. Strategic Decision Corp. accomplishes this by combining leading-edge technology and customer analysis into a single package. Strategic Decision Corp.'s product suite, PersonalClerk, merges five key consumer data analysis capabilities:

- **Segmentation of Customers**—Our exclusive segmentation system, ZipSegments – built on the wealth of demographic data in the U.S. Census and triggered by ZIP code – allows clients to target promotions and products to appropriate customer groups, increasing return on marketing investment.
- **Knowledge Management**—Clients can view and easily understand customer data and design appropriate marketing programs to capitalize on this learning. Competitive technologies, such as collaborative filtering and neural nets, are "black boxes" where the underlying meaning is difficult to discern.
- **Robust Data Analysis Technology**—Strategic Decision Corp.'s patent-pending datamining algorithms support real-time updating of the customer knowledge base, as compared with traditional (neural net) datamining techniques that require repeated time-intensive batch processing of large databases.
- **Real-time Marketing Action**—PersonalClerk generates rules appropriate to each consumer in real time, allowing personalized online and offline marketing messages for each consumer, including up-to-the-minute content, promotions, and offers.
- **Security**—Strategic Decision Corp. supports PersonalClerk remotely, leaving all individual customer data securely within our clients' private networks.

PersonalClerk analyzes all available customer data, including location (either through explicit ZIP code inputs or conclusions from IP addresses), registration and explicit inputs, clickstream, and transaction history. In addition, Strategic Decision Corp. provides complete installation and ongoing support services. These services include integration with existing client systems and databases, ongoing updates to the software and ZipSegments database, and assistance managing rules and promotions.

> Since Strategic Decision Corp. was formed in November 1999, the firm has moved quickly to begin serving clients. The original architects of the company's services and technologies, Lee Cooper (professor of marketing at UCLA) and Giovanni Giuffrida (PhD candidate, UCLA Computer Science), filled the chairman and chief technology officer positions. Experienced marketing, technology, and consulting professionals rapidly filled other key management positions.
>
> Strategic Decision Corp.'s strategy is to rapidly expand the firm's client base. Beyond the e-retailer market, the firm plans to expand into Internet advertising and e-malls, where the segmentation and cross-selling capabilities of PersonalClerk can be best utilized. The company is expected to grow rapidly to fill the need for these products and services, resulting in compound annual revenue growth of over 355% from the end of year 1 through the end of year 4, reaching profitability after only 15 months in operation.
>
> Strategic Decision Corp. is currently seeking additional funding in order to launch a major client acquisition effort and continue our product development. This funding will allow the firm to demonstrate the advantages of the PersonalClerk product to a wide audience and capture significant revenues and growth, thus maximizing value for investors.

In the original document this executive summary fills one page, a practical maximum. Not everyone will read on; summaries are supposed to partition an audience into those interested in knowing more and those not. Our goals in the summary were to cover the four risks and be clear and compelling enough that the audience members we ultimately wanted were among those who chose to continue. The body of the plan, discussed in Chapter 6, had to deliver on the promise of the executive summary.

3.4 The First Meeting of the Board of Directors

On February 4, 2000, we held the first board of directors meeting in Skip's main conference room. I had already sent everyone an agenda and the business plan as it then stood. Steve and Penny flew in from San Francisco and Chicago, respectively. I invited Fred Hart to observe. He thanked me at the meeting, and said sincerely that this was the best business plan he'd ever read. When a billionaire says this to you, it's a somewhat heady experience. One must be careful not to presume anything from such a statement other than temporal flattery. This in itself doesn't signal long-term (or even medium-term) commitment to building the underlying business. You are probably one of 50 investments in a portfolio. The objective function is a combination of the VC criterion (i.e., return on partners' time) and the personal hedonic return (i.e., "Are these people fun to be around?"). Still, it felt good.

Chatting with Fred and Steve before the meeting got under way, I got another lesson in perspective. Fred said, "Hollywood is great. You can make hundreds of millions of dollars. But you'll never make *billions*." My takeaway from his statement was simple: Of the billionaires on this planet, all but one have someone richer to envy.

We began by introducing all of the staff to the board. We followed with a demonstration of the prototype. At that point all the staff left except Jason, Van, and the attorneys. I noted the progress to date, and then shifted to discussing potential beta sites: We were to meet with Guidance Solutions the following Tuesday. Burt Richards, a candidate for the VP sales slot, set up contacts with Toy Time, Casino Play, and Artists Direct. In the small world of Hollywood, at least three others in the room had major connections with Artists Direct that could be used at the right moment. eHobbies was available as a beta site, but it was hosted in Northern California.

Fred expressed concern about sales force that I tried to address by citing Burt Richards's experience and his plans to bring in three additional people.[35] Bryce wanted to know how soon a beta would be ready, and I indicated we could be ready a week after the five development platforms IBM promised to loan us were delivered. Fred raised the question of when more money would be needed; I answered early April, depending on how aggressively we grew. The extended discussion of funding that followed is developed in Chapter 6, "Smart Money."

Alliance development was next: We were passed the first screen A.T. Kearney, and Jason's contact provided a champion high in the organization. A.T. Kearney has three screens, the last of which comes with money. eMarketing Partners was very interested, awaiting demo,

---

[35] I didn't know then that we would never convince Burt Richards to sign on with us.

as was the KPMG Digital Media Incubator that Mariana Danilovic, a former UCLA MBA student, set up. Meetings with Viant and Etinicity were being set up. Burt Richards had a three-facet alliance plan for consulting firms, digital agencies, and integrators.

We discussed client contacts: The somewhat frustrating experience of our conference call with Fingerhut occurred the morning after it announced an agreement with Net Perceptions. This frustration continued the next week when we met with Guidance Solutions' CTO and VP of engineering immediately after a Net Perceptions team. We were facing competition that was spending 110% of revenue on marketing and sales. We needed to demonstrate product leadership in an arena where rumors and promises were ubiquitous, and result elusive.

We decided to add Martin Ross and Daniel Favor to our advisory panel, partially for their expertise and partly to compensate them for their efforts in due diligence. Professor Daniel Favor is a chaired professor of Telecommunication Systems at a very major east coast university. He focuses on leading-edge research in high-speed networking. He is a wonderful avuncular character whom Len Steiner had arranged for me to meet to discuss network security designs. Billed as an informal chat while on his way to a Christmas vacation in Hawaii, this all such meetings served multiple purposes. I learned our basic structure was sound, and gathered some tips I passed on to Giovanni for improving it. Apparently we passed our side of this due diligence. Shortly after our meeting, Dr. Favor was appointed to a top advisory role with the Federal Communications Commission.

By the end of our first meeting, the board members were enthusiastic and ready to use their fat Rolodexes to further the interests of the company. What more could I want?

That night Ann and I celebrated with Steve, Penny, and Bryce and his wife Kathy. Bryce was amazed that the summer sailing fantasies were now reality. Steve and Penny, having known me longer, may have been even more surprised. Was I surprised? Yes. I knew the odds were very much against making it this far. I knew this was a milestone worth celebrating. I also knew I had a hold of a compelling and scalable idea that could help make e-commerce as profitable as everyone involved hoped.

4. I Should Have Read Charles Ferguson

This chapter relates all the steps and missteps in the search for a permanent CEO -- and the consequence of our final choice.

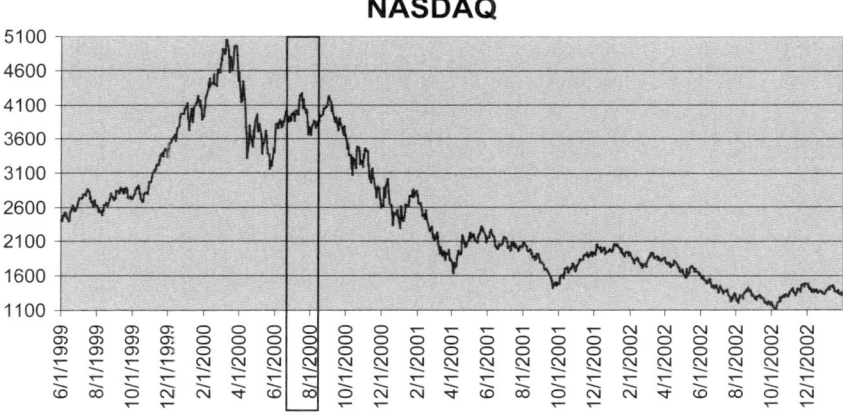

4.1 Whitewater Canoeing

I had met Ed Pinter through my neighborhood, and got to know him better during long discussions at our local Boy Scout troop's Family Camp held each Memorial Day weekend. His heavy travel schedule kept him from most weekend campouts, and he had never been on the high-adventure weeks during the summers. But Family Camp was a more civilized experience, when wife Pat and daughter Margaret could join Ed, his son Johnny, and all the neighborhood scouting families that made the trek to a camping site on a lake near Paso Robles. The parents watched while the scouts participated in canoe regattas, practicing in sometimes silly-looking competitive games the skills that could save their lives in high adventures. In the times between regattas, wide-games, or campfires, however, the parents gathered for social talks. Bryce, another neighbor I initially met through the Boy Scout troop, and Ed were two of my favorite people to chat with long before this venture germinated. With Ed, the topics centered on current events, international affairs, markets, and captivating stories of dealing with Philippine presidents, Australian premiers, Saudi princes, and employees stuck in Saudi jails. I knew he was an undergrad math major at Dartmouth and a Yale-trained attorney, who was CEO of the energy-development arm of a major utility. I could tell he loved the art of the deal. Making deals happen added sparkle to his eyes.

In the week following the first board meeting, Bryce told me that Ed was available, and I should try to get him interested in this venture. Intuitively, I completely agreed. I gave Ed the general idea over the phone and arranged to show him the demo on February 2, the day after the Galef Symposium on Corporate Governance at UCLA. Before the demo I asked for Ed's résumé, from which I learned that in his six years at the helm of a major independent power producer he had grown the earnings (real earnings!) from $2 million to $198 million, and in the prior two years had created and financed $8.5 billion in energy projects. He had great credibility with investment bankers and analysts in the energy arena, but he pointed out that these were not the same parties that followed and funded the Internet and e-commerce.

From the demo, he quickly grasped what we were trying to do. I already knew that he had the poise and savvy to handle the CEO-to-CEO contacts we anticipated would be needed. The only question in both our minds was how quickly he could come up to speed on what was happening in the Internet economy. He didn't try to sell his knowledge in that arena, emphasizing his experience has been in what he characterized as *rust-belt business*. Still, I believed that Ed was the right guy, and I decided to introduce him to the rest of the board with the CEO slot in mind. Paul made the call that reserved the private dining room at 346 Maple Drive for the following Monday. Allying the company with Hollywood heavyweights did have some advantages. Steve flew down from San Francisco, but Penny met Ed via phone on Tuesday morning. Due to the quick scheduling, we had to set up separate meetings with Bud and Fred. So, Steve, Paul, Edward Lennon, Bryce, Ed, and I enjoyed some great food and conversation. Bryce was already in Ed's corner, but added one more known face to the assembly. The conversation was wide-ranging and subtle. I tried to listen to the agendas being worked. This kind of dinner was not my normal academic fare.

The feedback I gathered later was that Ed was a very impressive gentleman and businessman who needed to catch up on the speed and inner workings of Internet companies – basically what we already knew. Bud was very impressed, as was Fred after separate meetings, but the issue we didn't try to gloss over was always present. Ed lacked *Internet DNA* – that supposed inner sense of how things were done in the wired world. This criticism was leveled, perhaps unfairly, at the rest of the company management, though Jason had been an early Netizen, Van was from the networked world, and I had been online with email since the BITNET days of the early 1980s. Conducting research with colleagues in Japan, Australia, France, and Germany was feasible only because of early electronic networks. But I hadn't done business deals in this strange, wired world. So as not to appear defensive, I listened to what others claimed was our corporate flat side.

Ed believed in building businesses with real earnings. In retrospect, the basic business perspective that he brought made a lot more long-term sense than the go-go, bubble logic of the times. Nonetheless, the consensus developed that we should find a way to get Ed more involved in the venture and see how rapidly he climbed the knowledge curve. I talked Ed into joining the advisory panel, and started using him actively as a sounding board for issues that troubled me. Much to his credit, he was happy to help. He joined our early discussion with Pennyweb, a cost-per-click banner-ad network, and helped me articulate my vague feeling of ill ease into a rational balance sheet that revealed we were bringing a lot more to the table in terms of intellectual capital than we could ever get from a valuation standpoint. It was not the time to do a deal.

I continually learned from my interaction with Ed, and valued the relationship greatly. While I held unfruitful discussions with Andrew Harper, who professed reluctance to re-engage in an active management role so shortly after leaving HouseFinder.com, I grew more convinced that Ed was the right guy for the CEO role. Whitewater canoeing together would test that conviction. I needed to know if I could follow his lead as well as if he could follow mine. The kind of partnership I hoped for between me and a CEO was one that required mutual trust and shared leadership.

You're probably imagining some Outward Bound, male-bonding experience. That's not very close to the truth. I had been encouraging Ed to take the time, since it was finally available, to be more a part of his son's Boy Scout experience. For the prior nine years, I had gone to at least one week of the summer high-adventure experience that the troop sets up, the first six with my older son, one year of overlap when both boys were in the troop, and the last two with just my younger son. I told Ed what a wonderful experience it was for me to watch a child grow in these annual snapshots. Our day-to-day experience with kids lacks the discrete markers that can be more vivid reminders of progress. Like a slow-ticking clock, each annual adventure becomes a vivid moment, isolated in time, by which the miraculous transformation from boy to young man is chronicled.

In mid-August 2000, the plan was to fly up to Oregon and drive down to the Klamath River, joining our sons who were to be bussed up after a first week at a Boy Scout camp at Shaver Lake. Five days of canoeing and camping were to follow. Only the last day would fathers and sons canoe as pairs; the rest of the time the scouts would pair up and the dads would do the same. Ed had canoed extensively in his rural-New Jersey youth. I had a lot of experience in recent years canoeing with the scouts. Ed and I planned to partner. My hidden agenda was to see if I could follow his lead when he was coxswain, and if he could follow mine when the roles were reversed. I know my tendency was to steer regardless of being in the front or the back. In the early days of the river trip we alternated positions in the canoe. When I was coxswain, Ed executed his supporting role without conflict. When I was in the bow, I found out that for the most part I saw the river the same way he did, and had no problem following his directions. By the time we ran into rapids where I was unsure, or even had a different opinion of how we should proceed, I had developed enough trust in his skills to follow his lead. There are many ways to solve problems. His way proved to be as effective as I imagined my alternative would be. I learned we were comfortable in either role, and I was taught an even more important lesson on the penultimate day.

Whitewater canoeing can be quite daunting. Class 2 and 3 rapids that pose little threat in the mammoth rubber rafts can be treacherous in an open canoe. Upriver of each major rapid, we would gather the canoes and scout the prospects on foot. Our experienced guides instructed scouts and dads on paths through the obvious and not-so-obvious hazards. Canoes would venture one at a time, waiting for an all-clear signal before proceeding. While on many rapids some pairs swamped or capsized, the first three days contained much joy and no serious injury. But the Class 3 rapid we faced on that fourth day looked scary.

A large, silent pond converged into a smooth-rocked outlet with only one narrow channel deep enough for clear paddling. Over the first fall on the right was a relatively clear sweep of water that would force the canoe into a vicious pinnacle rock nearer the bottom of the second fall. Midstream lay a partially exposed boulder forming the outer lip of a barely submerged ridge of the second fall. The first fall and this ridge formed a backwater trap. The safe path required two maneuvers: The bowman needed to sweep on the starboard side while the coxswain J-stroked on the port side until the canoe was past the threat of being drawn into the clearing on the right with the pinnacle rock below. The bowman then had to draw on the starboard, while the coxswain swept on port to pass to the right of the midstream boulder. This would position the canoe to pass to the left of the pinnacle rock and be centered enough to follow the rushing white water without crashing into the shear rock wall on the right side of the stream below.

I stood on the right above the first fall, relaying the all-clear signal to the waiting line of canoes. Each time rescuers corralled the capsized canoes, and scouts and dads were helped to the safety of the lee on the left far downstream, I would pass on the *go* signal. Some of the older scouts, expert and fearless, made it through fine. Everyone so far avoided the pinnacle rock, but most ended up capsized in the stream. My son swamped, but stayed with the canoe until helpers pulled it to safety. The growing group of survivors hiked back on the left side of the rapids to watch the canoes come through.

As I walked back to our canoe, I visualized my role as coxswain: J-stroke past the first hazard, sweep past the second boulder, and then J-stroke to stay clear of the wall. We approached the rapid paddling slowly, Ed starboard and me port. The current over the first fall was very swift and I needed strong steering to stay left. When we cleared the first hazard and I tried to sweep right of the boulder, I knew my strongest sweep would only send us directly into it. I pivoted the canoe into the backwater trap and we found ourselves facing upstream into the first fall, standing still amid the turbulent water

180° out of phase. I was about to try a reverse draw to pull us around the boulder and out of the trap when one of the guides yelled from the bank, "Paddle forward." Well, "forward" was a raging waterfall that even a determined salmon might not surmount. "Forward" was paddling against a tremendously strong current. Nonetheless, paddle forward is what we did. Strong strokes inched the canoe closer to the first fall, catching the strong current running midstream of the boulder and, still paddling slowly forward, moving us backward smoothly, under control, around the boulder. Facing upstream the whole way, we navigated downstream, clear of the pinnacle rock and away from the wall, to the cheers and laughter of the rest of our group.

Both Ed and I knew that we had had a remarkable experience. The lasting lesson from that experience didn't become clear to me until too many months later. We faced currents so strong we could not overpower them. But unless we were willing to paddle against those insurmountable forces, we could not expect to align our efforts to get where we wanted to go. Paddling against the flow provided the control our canoe needed, and slowed down the rush so we could see how to navigate.

4.2   Return to the Real World

One of the ongoing problems of residence-based learning programs – be they in the wilderness, the luxurious retreat center, or on the university campus – concerns integrating what the participants learn into their work or personal lives. Ed and I left the river with a plan to form an Office of the Chairman, with Ed, Andrew, and me working as a team to push the company forward. All we had to do was convince Andrew to play his role in this triumvirate. We returned from the river on August 19, and I set a meeting with Andrew for Tuesday, August 22.

Monday Paul Brendl came over after our 9 a.m. staff meeting. We needed to coordinate before our 3 p.m. meeting with Farhad Mohit, the chairman and chief strategic officer of BizRate.com. BizRate.com was taking the next steps in the transformation from the e-commerce rating service, which provided its practical start, to the e-commerce portal that had always been central to Mohit's vision. A recommendation engine such as PersonalClerk fit ideally with that vision, in our humble opinion. Convincing BizRate.com to use our recommendation engine and value-pricing approach was our agenda. Around 10:15, Giovanni called to say we were about to throw the switch from learning to optimized execution in the iPlayer.com ad test.

Remember back to the opening episode of the book. July 28 we had begun the learning phase of our ad test. While in normal execution each 30-day ad would take a day or two to accumulate sufficient learning to begin optimized execution, in the test we were getting very little traffic directed to us and we had to learn on 215 ads all at once. For this test we had initially allocated 15 million impressions for learning, and 5 million for optimized execution. When the final ad count went up to 215, I figured we needed to increase ad learning to 20 million (with five reserved for execution), use the 22-segment version of ZipSegments, and drop the 12 age categories. I believed the 68-segment version was the best scheme, but with three genders (male, female, and unknown), we would only have an average of 350 impressions per ad for learning. With click rates so low, this wasn't enough to let us use the 68-segment system. But ZipSegments was designed with this practicality in mind. It is a hierarchical system that collapses the 68-cluster system into larger clusters that can be used depending on the amount of data available. The learning algorithm simultaneously gathers evidence at the 68-, 45-, 22-, and 11-segment levels. It also automatically learns with and without age and gender distinctions. If we used the 22-segment system, and collapsed the age categories, we would average more than 1,000 impressions per banner ad per segment-gender combination – a much sounder basis for estimating what were the most preferred ads.

The 20 millionth impression was about to be served. Paul and I went down to the second floor office where Giovanni, Fabrizio, and Giuseppe watched a log screen reflecting impressions shown and clicks recorded. The baseline click rate among registered users was slightly more than two per thousand, 0.21%, close to what the industry had been learning to expect. As we flipped the algorithm over to execution, we started feeding each segment what we had learned it liked most to click, rotating among the most popular to reduce wear-out. Almost instantly we could see the click rate increase. Within minutes it started to level out at 0.83%, four times the base rate. We knew then as we know now that click rate isn't the ultimate criterion for advertising effectiveness, but a 4× lift[36] in any ad-effectiveness index as a result of understandable targeting is the kind of result that people notice. It would mean the quadrupling of effective inventory for any banner-ad supported Web site.

Still, something wasn't right. My rough estimates, based on simple extrapolations of click rates we observed during learning, indicated the results might be as much as twice the lift the log screen showed – and overall, 7× to 8× lift. Simulations using the best 10 ads in each segment-gender gave more than a 12× lift, and using the best 20 ads in each segment-gender cell gave a 9× lift. I knew my estimates didn't weight wear-out, but still I expected more than the current 4× lift. I asked Virginia Eastwood, a doctoral student in statistics at UCLA who worked part time at SDC, to get estimates of the click rates for the top ads in each segment-gender combination.

Paul and I left for the meeting with BizRate's Farhad Mohit and the vice president of data systems and operations. After two meetings with technical teams where we sold our technological capabilities, this was our first meeting with BizRate.com decision makers and deal shapers. They were interested in a code escrow to guard against our company disappearing, and most-favored-nation (MFN) status. This would automatically revise their deal in line with any better deal that anyone else received. We were listening rather than negotiating at that point, or Paul would have opposed MFN on the spot. No two deals are exactly the same and such clauses end up in arguments over pieces of a total package. They also wanted the ability to create proprietary versions that extended our technology. George Rebane, VP of advanced projects who earned his PhD under Judea Pearl at UCLA many years before, was working on Bayesian approaches to recommendation engines that he thought would complement our efforts. That was the Party Line. My sense was that George recognized that we were tactically far ahead on the customer-facing side, since he and his partner Martin Schmidt, VP of engineering, had focused mostly on advanced database structure and search logic. George had plans for integrating our developments with his Bayesian notions. What we liked about this was that it fed directly into our desire for value-based pricing. We wanted to get a slice of the enhanced performance we provided rather than a flat fee. Getting a highly visible client such as BizRate.com to buy into this would make

---

[36] Lift is a measure of increase over a baseline performance. A 4× lift reflects four times the baseline performance.

it easier to persuade later clients. It required a small experimental design in which some baseline performance was monitored of his approach alone, our approach alone, and the combination. The devil is in the details that were left to later discussions, but Paul and I left with the clear sense they wanted to do the deal.

On returning late to SDC I found that Virginia's report wasn't ready. Our execution click rate was 0.81% after 610,000 impressions. I didn't want to use all our ads before knowing we were optimizing correctly.

The next morning, we awaited word from a client-services team we sent off to Little Rock to meet with Acxiom, the largest consumer-data firm in the world. Acxiom's client list read like a Who's-Who of the top 100 retail companies. The company worked with 24 of the top 25 credit-card companies. It planned to launch AbiliTec as a major Web effort and wanted to evaluate our recommendation engine as a key component.

At my 10 a.m. meeting with Andrew I shared the current execution click rate along with my concerns that we might not yet have it right, and I broached the idea of the Office of the Chairman. He was noncommittal and cool to the concept, wanting to wait until more results from the ad test were available. By end of day I found out that the baseline for comparison had actually dropped over the week before we had thrown the switch to 0.18%. We had over 4.5× lift, but my uncertainty remained.

In the afternoon, Bud called for an update on the test and to discuss his strategy. He wanted to do deals in three areas: wireless, ad agencies, and portals. He spoke in broad strokes, but I knew the key drivers of success in each area. The wireless play was clear. As the visual geography becomes more scarce, good recommendations become more important. You may have only one crack at getting it right. For ad agencies, the stakes were high. The Internet held the promise of demonstrable ad effectiveness – the ultimate measurable environment. Since the segmentation scheme could be ported from the Net to direct mail and to media buys, agencies could use us to integrate their efforts across marketing channels. The major portals, MSN, AOL, and Yahoo!, were striving for increased ad and e-commerce revenue. The recommendation engine and ad-optimization scheme worked together to turn these portals into virtual communities and virtual malls with an end-to-end personalized experience – the realization of the theory in the Hagel and Armstrong (1997) vision of life on the Net.[37] Bud thought that, armed with the ad-test results as a proof of concept, we needed only a senior sales exec and a CFO. I would run R & D, and our current team could run sales and customer support. I talked about the Office of the Chairman, but hardly got a word in amid Bud's stream of enthusiasm.

By the end of the day the report still wasn't ready.

I spent Most of Wednesday at UCLA. We were recruiting Andrew Ainslie from the Cornell faculty, an empirical marketing scientist who complemented UCLA's growing dominance in this arena. Anand Bodapati, Bart Bronnenberg, Randy Bucklin, Mike Hanssens, and Don Morrison – these are the kind of scholars who take on the challenge of new data streams to enable firms to manage in information-rich environments. I got back to Santa Monica in time for a 4 p.m. call with Marc Singer, the co-author on John Hagel's second book.[38] We were supposed to talk about prospects for putting SDC into McKinsey's Accelerator. I, frankly, wasn't sure we could accelerate any faster. Given the number of times he had rescheduled this call, he might not have had room for us anyway.

End of that day, the reports still were not ready.

Thursday was another huge day. David Midgley, a member of our advisory panel, was in town from INSEAD, one of the leading French business schools. I invited him to join Ed Pinter, me, and much of our senior management to listen to a report on B-2-B opportunities for SDC. David's and Ed's knowledge of business-to-business networks helped me see the mismatch between our business model and the opportunities in this area presented by an Anderson School intern who had done an outstanding job of presenting his case. The Verticalnets of the world were flying high at the time with a split-adjusted stock price around $500, but Ed underscored how temporal this success would be since dyadic supply-chain relations would pull out of cooperatives as the partnerships matured. As I

---

[37] Hagel, John III and Arthur G. Armstrong (1997), *Net Gain: Expanding Markets through Virtual Communites*, Boston: Harvard Business School Publishing.
[38] John Hagel, John III and Marc Singer (1998), *Net Worth: Shaping Markets when Customers Make the Rules*, Boston: Harvard Business School Publishing.

write this, Verticalnet sells for $.67. At 3 p.m., we delayed our normal VPs meeting to confer with Alex Cohen, then CTO of POP.com. Alex was the director of technology at Netscape and created the lightweight directory access protocol (LDAP) that made the customization in *my.Netscape* possible. With roles as head of content technology at CNET and chief architect of advertising technology at Excite, learning to drink from a fire hydrant seemed to be his specialty. Bud had connected us, and our phone conversations convinced me that I needed help in our meeting to keep up with the pace of his thinking and speech. Giovanni and Ravi joined in, while Jason and a group from Client Services went over to BizRate.com for an integration meeting with its VP of engineering, Martin Schmidt.

End of the day Thursday, the report still was not ready. We were starting to run through 1.5 million ads per day. Friday might be the last full day. Our budget was certain to be exhausted over the weekend. Ravi and Giovanni were already overloaded with the iPlayer.com test, but I let them know that if Virginia couldn't get the reports I needed, they had to jump in. Their approach was to make sure Virginia got it right.

Friday: 12:46 p.m., I finally got an email from Virginia saying that the queries gave her multiple records for each segment, gender and banner ad. She should have a single click rate for each of the top 10 ads within these segment-gender combinations.

By 4:30 p.m., we had hard-to-decipher SQL output showing that ads not in the top 10 in each segment-gender combo were being served. At 5:13 p.m. an Excel report showed clearly the wrong ads were being served. Giovanni, Fabrizio, and Ravi had been combing the systems for problems all week. We knew all technical systems were operating properly. This had to be a command instruction that was not set as I had directed, and only two possible screw-ups came to mind. I stormed down to the second floor. Were we possibly using 68 ZipSegments rather than 22? No. Were we using the 12 age categories, rather than collapsing across ages? Yes! I tried to remain calm, but everyone knew how pissed I was. Using age for learning required 12 times as much data as we had. Determining the best ads on such sparse data was bound to lead to sub-optimal performance. Fabrizio changed one line in the command file and we began to execute on the more robust preferences learned by segment and gender across all age groups. We stood around the monitor in

Giovanni's room as the click rate increased: 0.8%, 0.9%, 1.0%, 1.1%, and 1.2%. Within 15 minutes the rate climbed to 1.22% – a 6.9× lift.

4.3   Irrational Exuberance

The technology team members celebrated the victory, and rapidly recovered from the thought that they almost blew the test. They wanted to know how soon we could halt the test so that they could use the few impressions that remained over the weekend to experiment with more efficient serving approaches that had occurred to them during the prior month. I wanted as many optimized impressions as I could get. We settled on shifting to experimental mode on Sunday. As word of the seven-fold lift began to filter out to the investors on the board late Friday and over the weekend, whispered irrational exuberance took hold. The naïve computation ran something like this: we could give a site the vast majority of the increased revenue from 1× lift to 2× lift. Since this would be incremental revenue with little or no incremental cost, it would drop straight to our client's bottom line. From 2× lift to 3× lift we could take a higher percent, and above 3× lift we could demand a 50-50 revenue split.

Remember, this is August 2000, and despite the acknowledged decline in e-commerce retail, Jupiter Communications was predicting, "Online advertising is poised for dramatic growth, driven by marketers' need to understand the efficacy of their advertising campaigns and the greater accountability that the Internet affords them." Jupiter forecast that online ad spending would grow to $16.5 billion by 2005.[39] Our approach was squarely aligned with marketers' needs, and boosted advertising effectiveness possibly up to sevenfold! Given the concentration of ad revenue in the top properties, the result of this expectation, computed across just the properties on the Net where Bud could access the key decision makers, amounted to billions, with gross margins over 95%.

I started hearing phrases like "low-hanging fruit" to justify a push toward the advertising side. Andrew, to prepare for giving feedback to iPlayer.com, started demanding simulations that imposed budget constraints that were relevant to the advertising side only where ad

---

[39] Jupiter Vision Report, "Online Advertising Through 2005: Flourishing in the Dot-com Decline," August, 2000.

buys were on a cost-per-thousand (CPM) basis and lasted one month or so. The retail recommendation suite, PersonalClerk, could expect the same range of lift without requiring severe budget constraints, but I kept being pushed to come up with different constrained scenarios. Eric Bradlow came into town from Wharton to help with the simulator. We found that if you collected the banner ads for the 49 different advertisers and used our results merely to redistribute each advertiser's budget among the banner ads it used, we would get 5.3× lift. Among advertisers with four or more banner ads, we achieved 8.2× lift by redistributing their fixed total budget.

When we fed back the test results to iPlayer.com, I could hear the company's CEO cranking through the calculations. If he could monetize the lift, he could buy related Net properties based on iPlayer.com's current advertising revenue and crank four or more times the revenue based on the company's using our technology. Separately, Fred started talking about a new $20-million round of financing at a $200 million pre-money valuation. By the end of the week, the idea of the Office of the Chairman was dead, and Andrew was slated to be the CEO, promising to grow SDC into a billion-dollar company within 18 months.

Time out! I had just come off the river with a clear understanding of how I could partner with Ed. I had learned that when facing an insurmountable current I still had to paddle against the stream to align my effort. Yet, when sucked into a trap much like the pool Ed and I faced on the Klamath River, I failed even to stroke against this onslaught. I started the previous section by noting that one of the ongoing problems of residence-based learning programs concerns integrating what the participants learn into their real-world lives. Obviously, I'm no exception.

I should have read Charles Ferguson's seriously good book, *High Stakes, No Prisoners: A Winner's Tale of Greed and Glory in the Internet Wars*.[40] My excuse: I had read Jerry Kaplan's book, *Start Up* – a cute if self-serving tale of Go Computers and the handheld OS that arrived before its time.[41] Save me from books written by people who need to get back in the game. I had also read Michael Wolff's book, *Burnrate* –

---

[40] Ferguson, Charles H. (1999), *High Stakes, No Prisoners: A Winner's Tale of Greed and Glory in the Internet Wars*. New York: Times Business – Random House.
[41] Kaplan, Jerry (1995), *Startup : A Silicon Valley Adventure*. Boston: Houghton Mifflin.

74 | Midlife ~~Crisis~~ Startup

a cautionary tale, very well told, of a good man's attempt to lie and cheat his way into the time and money needed to succeed.[42] So, when *High Stakes* came out, I thought of it as possibly redundant. I was wrong. Among the many lessons in that book was the clearly dramatized tale of choosing a CEO without adequate due diligence. Ferguson, whose CEO was pushed on him by certain board members, caught his CEO in lies and misrepresentations that were basically aimed at cutting a much better deal for the CEO than his contribution and short tenure merited. Were it not for the quick sale of Vermeer Technologies to Microsoft, the consequences of Ferguson's CEO choice could have been much more severe.

4.4   Due Diligence

What did I know about Andrew Harper? I knew he was a football lineman for one year at Navy, before transferring to UCLA. That position fit his personality much better than I then knew. At UCLA he studied economics, math, and computer science. After graduation he got a CPA, and worked for a Big-Eight accounting firm in its insurance practice. He was introduced to me as a two-time successful entrepreneur --having taken both startups public -- who took the B-Round board seat after investing $500,000. His first company was Clearview Information Systems, which created software that automated some of the insurance functions. I assume he left some time before 1994 when he started his second company. I didn't do the research that would have told me that Clearview Information Systems became ECIS. At the time I should have been thoroughly investigating his background, the SEC had just presented ECIS with a formal Order of Investigation. Quoting from the ECIS 10K filed in April, 2002 (page 8):

> We could be subject to civil fines and penalties as a result of the SEC's investigation of our financial reporting.
>
> On August 11, 2000, we were advised that the SEC had issued a formal Order of Investigation and subpoenaed documents relating to our financial reporting since April 1, 1997, including, in particular, revenue recognition, software development, cost capitalization, royalty costs and

---

[42] Wolff, Michael (1998), *Burn Rate: How I Survived the Gold Rush Years on the Internet.* Simon & Schuster.

> classification of cash receipts. We have submitted documents to the SEC upon the SEC's request as part of the investigation. It is possible that the SEC could impose civil fines and penalties against us. An adverse finding against us by the SEC could negatively impact our stock price. In addition, we expect to continue to incur expenses associated with responding to this investigation, regardless of its outcome, and this investigation may divert the efforts and attention of our management team from normal business operations.

I assume Andrew was long gone by April 1, 1997. But he was a CPA who was the co-founder and CEO. Who set up the accounting policies? I should have found out.

Andrew told me that in 1994 he founded HouseFinder, a touch-screen information kiosk that provided residential real-estate listings. In November of 1995, HouseFinder negotiated an exclusive revenue-sharing arrangement with the American Association of Realtors. According to Andrew, in 1996 some unspecified mess arose that led to a White Knight coming in as CEO, with Andrew as president. HouseFinder ultimately became HouseFinder.com and went public in August 1999. What I found out later was that the White Knight was brought in as CEO at the insistence of Kleiner Perkins Caufield & Byers. KPCB usually enters investments at the A or B Round, as they had done with Steve Mayer in the Digital F/X deal. At HouseFinder.com they entered in a D Round, forcing Andrew to take a demotion from number one to the number-two slot. The White Knight was a PhD in electrical engineering with prior experience as vice president of business services at two very large subsidiaries of a major communications conglomerate. Andrew left the board of HouseFinder.com at the end of 2000, reportedly over the restrictions officers of a company have in selling stock. He wanted to divest his HouseFinder.com holdings faster than the SEC would otherwise allow. Since then, of course, we have witnessed the high-profile stories about round tripping and misstatements of revenue at HouseFinder.com going back to FY 2000, so far. The White Knight resigned. Andrew's successor as HouseFinder's chief operating officer, and the former chief financial officer pleaded guilty to federal criminal charges of conspiracy to commit securities fraud. The CFO also pleaded guilty to wire fraud. They are cooperating with the ongoing investigations of HouseFinder.

The initial S-1 filed in May of 1999, prior to HouseFinder.com's IPO, showed Andrew owned 700,000 shares of common stock. That was upped to 1,750,000 shares in the S-1A after the SEC required amendments to the S-1. He had already received $4.3 million in cash for selling his HouseFinder shares to NetSelect before it transformed into HouseFinder.com. In the S1, we found that "On August 23, 1994, the company entered into a loan and security agreement with Andrew Harper, its founder, president and chief executive officer. The agreement calls for Mr. Harper to make loans to the company at a monthly interest rate of 10%." The loans were due and payable on August 23, 1995 and were listed as due on demand in the original S-1. The company owed him $150,000 by the end of October 1996. The initial loan amount wasn't disclosed, but loaning less than $13,000 to the company on these terms was all the investment that was needed to produce the $150,000 by October. Not a bad deal for Andrew. If he just let that loan sit until the IPO, the compounding interest would generate more than $3.5 million in debt owed Andrew. From $13,000 to $3.5 million is the kind of deal that might make the Sopranos blush. I assume, without evidence, that the loan balance was what translated 700K shares in the original S-1 into the 1.7 million shares Andrew owned at IPO. If this is the source, then what might have started as a $13K loan translated into more than $22 million at the IPO price.

I probably could not have found out about the SEC investigations of accounting practices at both of his previous companies. The other information, though, could have been gathered in less than a day from publicly available resources. If I had put together this profile, I would have had the information I needed to defend a decision against putting Andrew into the CEO slot. Without this profile I wasn't inclined to protest. The Fall Quarter began in late September. I felt I had to get a CEO in place before then, or my UCLA career would be at risk.

As it stood, Bud had already written the scenario: *Two-time successful entrepreneur leads SDC to billion-dollar IPO.* When friction later built to crisis level, Fred privately advised me to just lean back and let Andrew make me rich. Bud and Fred were engaging in Hollywood-like thinking. I was cast as the dissident writer upset with how the producer or director was modifying my script. Such thinking comes from the playbook of venture capitalists such as Kleiner Perkins.

They are notorious for ousting founders from positions of power. Steve Mayer noted that he was the only founder not bounced by KPCB – a fact he attributed to his aversion to conflict and his acknowledgement that accepting KPCB money meant Digital F/X was no longer his company. For VCs the issues are power and control, when what they should focus on is maximizing shareholder wealth.

4.5   The Seeds of Conflict

Bud, Andrew, and I met at Bud's favorite lunch spot, La Cachette on Little Santa Monica Blvd., to set the final terms of Andrew's deal. I expected issues of control and authority to arise. I tried to position myself as managing director – analogous to the role Penny had taken in her company. Andrew bristled at the thought and I backed off. In the Series B negotiation 5% had been set aside as an option pool for the CEO, about 1 million shares. The *about* wasn't good enough for Andrew, since it didn't round in his favor. He played with the denominator until the 5% translated into 1,312,500 shares. The original idea was the CEO options would vest over four years, as did most other employee options. Andrew insisted on vesting over 18 months, and formally contracting for three days per week for the first year and two days per week thereafter. In the off-the-contract discussion Andrew indicated he was a workaholic who would put in whatever time it took to get the job done. That was good enough for Bud. Andrew was wealthy from his prior businesses so, when Andrew suggested that he work without salary, Bud was willing to concede almost anything else he wanted.

Strangely enough, one of the biggest problems for Bud was Andrew's desire to put another $1 million into more stock at the B-Round price. Bud refused to go back to all the investors, since they might ask for the same special deal. He pressured me (along with Giovanni) to sell founder's shares. We agreed on $3 per share, about one-third of the way between the B-Round price and the expected price in the next round. Neither Giovanni nor I ever saw a penny of the million dollars. Founder's shares were common shares, as were Andrew's options. If he paid $3 per share for our shares, then the market value for the options for which he was scheduled to pay $.20 per share would be $3 per share. That would leave him with an ordinary-income gain of around $4 million for an illiquid asset. Once he realized that, handshake agreements didn't matter. One of the many ways founders get screwed is by arbitrarily depressing the value of common shares in relation to preferred shares. Until a public market exists in the stock, few checks on accuracy of valuation are available. By claiming the low value as the fair-market value, any gain held long enough will be treated as a capital gain, rather than normal income.

So we had a CEO working without salary, who obviously believed in the potential of the company. What was the problem? The million dollars waiting to be invested would buy only a tiny slice in the next round, if Fred's idea of valuation held. The prospect that Andrew could keep the valuation low so that he could buy more seemed remote. These did not seem to be viable seeds of conflict.

Bud arranged for Goldman Sachs to come out to help value the firm. On September 14, Andrew and I met with two top people from the L.A. office and a top analyst from the Investment Banking Division of the Silicon Valley Office. The top LA person was already familiar with our operation through his prior contacts with Bud, but we faced tough questioning from the other two who held a JD and a PhD in economics, respectively, from Stanford. While we successfully responded to many business and technology questions, the breakthrough came when I made them understand how knowing *a little* with statistical certainty was better than knowing 800 pieces of unreliable data – which was how we characterized the Engage.com approach. You never know where due-diligence inquiry will take the conversation. In this case, I had to explain how the correlation of two variables, say a characteristic X of customers and preference Y for a particular product, can be no higher than the reliability with which you measure X.[43] Once they were satisfied that sound statistical principles backed up what they felt was a counterintuitive approach (i.e., less is more), we could see their enthusiasm for our undertaking grow. We heard back shortly that they wanted to invest in us, rather than just help us establish a valuation. The first word was that they wanted to lead the next round.

Further, Bud at this point was touting our test results to the President of AOL.[44]

> Dear Bob:

---

[43] Actually the Index of Reliability for X, which is the square root of the normally derived Reliability Coefficient, represents the maximum correlation between X and any Y.
[44] September 12, 2000, email.

> I want to tell you about a company named Strategic Decision Corp. that I believe can have a major impact on both AOL's revenues and profitability. SDC has just completed a large-scale suite of tests of its technology, which confirmed that it dramatically improves the effectiveness of Internet advertising (an improvement in click-through rates to 500% –700% of the rate prior to using SDC). SDC uses a sophisticated proprietary rules based real time data mining technology.
>
> There have been so many false promises and lies in this arena that I feel that I must say that this one isn't vaporware. This company truly can deliver. Further, I have personally known the founder and chairman for over 30 years, and besides being one of the world's most widely respected experts in marketing statistical analysis, he is also one of the most forthright and solid individuals I know. I believe that SDC's technology will have a dramatic effect on the advertising paradigm and economics of the Internet. Moreover, the technology also provides similar improvements in product recommendations and other promotions. SDC is prepared to immediately demonstrate their revolutionary technology.
>
> I would like to arrange a meeting with you and both SDC's CEO and its founder. If you can spare an hour to discuss the tremendous results they are getting, I think you will agree that it was time very well spent.
>
> I have included a brief summary of SDC below and will call you in a few days to discuss this matter.
>
> Thanks!
> Bud

A similar email went off to the head of Lycos, top contacts at NBCi, and associates with connections at the top of Softbank and Yahoo! If I send such emails to the head of the leading Internet portals, they get deleted without being read. Bud got a positive reply as soon as AOL's top guy returned from Europe.

On September 12 we combined an appearance at the Internet Commerce Exposition in San Jose with a meeting with AdForce, which at that time served the iPlayer.com ads, over possible integration and joint projects with other clients. After our basic presentation of test results, Andrew felt the need to assert his new position by impromptu diagramming of potential integration plans. I understood his desire to appear in control, but his technology vision scared the hell out of Giovanni and me. He didn't seem to understand how our optimization worked – basically sandwiching our technology within layers of AdForce in a way that minimized our potential to lift performance, and eliminated our ability to translate improved performance into business intelligence. What he wanted was an SDC utility that could plug in anywhere, delivering revenue to us each time it's called. While this is a cute fantasy as a business model, we should remember Clayton Christensen's warnings about radically new technology.

To repeat, Christensen asserts that innovating companies face a fundamental choice when commercializing a disruptive technology. Option 1 is to accept the market's needs as well defined and push the technology to its limit to address the needs of the existing market. Option 2 is to accept the technology's current capabilities as given and seek the market that will value the inherent attributes of that technology. Christensen asserts, and I agree, that Option 2 is the most successful route. What Andrew was advocating in the AdForce meeting was Option 1.

Between the beginning of September (when Andrew starting coming in regularly) and the middle of October we encountered a series of what I came to call the Friday Surprises.

The first occurred at a Friday (9/22) meeting on imposing business constraints on the optimization algorithm. The heart of the difference between the retail application of PersonalClerk and the AdServer model involves the way banner ads are sold to Web publishing sites such as iPlayer.com. An advertiser will pay X dollars for, say, 1 million impressions shown on the site during a month. If advertisers pay the money, they want the ad shown 1 million times. The advertisers will choose the sites on which they advertise by the demographics on the site (i.e., Is it the right target audience?) and the click rate (i.e., Did anybody notice?). How to impose these ad-budget constraints on the AdServer algorithm was the issue. What AdServer naïvely did was order ads in each segment by the probability that someone in a particular segment-gender combination would click on them. As people requested further pages on the site, the new ads delivered with the new page would be next on the list. This ensured that the ads with the highest click rate in each segment-gender combination would be shown the most. Andrew seemed to be pushing to come up with an approach based on suggestions of Jason

(Client Services) and Chas Frittoli (Sales). Given that Chas never had an algorithmic idea, I wonder if he and Jason were just fronting Andrew's notions. In any case, the notion was to form an index associated with how much trouble a particular campaign was in and multiply that index by the AdServer score to reorder ad delivery. The *trouble index* was based on a bunch of factors that were not measured well. The lack of a proper metric basis for any of a series of *factors* that were to be multiplied together meant the resulting product could be meaningless. This undisciplined approach puzzled Giovanni and seriously scared me. I pulled the proposal off the table, saying I'd consider the issues over the weekend. I spent the weekend drafting the fundamental equations of the business-rules algorithm. The basics were built on sums and counts that I knew would aggregate properly in the multi-level learning scheme that drove our approach. Monday I came in, gave Andrew the draft of the business-rules algorithm, and asked Andrew to spell out the problems that I needed to solve, not his answers.

The following Friday (9/29) morning, Andrew gave me his paper outlining his business-rules algorithm, wanting my critique. It was so naïve, I didn't know where to start — two pages, single-spaced, of definitions of mostly unmeasurable or unforecastable terms, but the definitions had equation signs in them so they looked almost like mathematics. These terms were combined into meaningless indices that supposedly could lead to a reordering of the ad priority lists for each cell. I spent half a day trying to figure out a way to respond. I finally pointed out a series of incomplete and inconsistent elements. I gave him a bottom line that even if our algorithms were equivalent, I would implement mine, since I was the one who had to solve the problems with it. Again, I emphasized that I needed him to spell out the problems, not his solutions.

The next week, Andrew gave me a memo that had a long list of the issues he felt the algorithm had to solve, along with some items that were algorithmic suggestions rather than questions. We worked all week to show how the algorithm I designed addressed each of his issues. His reaction in a Friday (10/6) private meeting was to berate me for purposefully ignoring his algorithmic suggestions. He accused me of treating him like a failed doctoral student. There was an element of truth in this. If Andrew was doctoral student, and had presented me with such ill-formed ideas and naïve mathematics, I would have failed him. Still, I wondered why Andrew was seeking that kind of validation. He wasn't a doctoral student, and he wasn't hired to be one.

That weekend was the first monthly camping trip for the Boy Scout troop. Paul Brendl's son had just joined, unaware that this was the troop with Ed's, Bryce's, and my sons. So Paul, Ed, and I talked about the company in general and Andrew specifically most of Saturday and on Sunday morning. It is hard for an outsider to come in as the new CEO of a startup. The culture is tacitly integrated into the work-group styles. SDC, in particular, was a very credentialed workforce. The technology team was heavily weighted with master's and doctoral degrees. The management team consisted of mostly MBAs. Andrew had neither. He was a CPA with a newly revealed penchant for micromanaging that seemed now very consonant with his early experience as an auditor. I understood Andrew's need to establish his position in the company. But I did not accept his need to alter the technology.

Sunday night was Erev Yom Kippur, and I was able to put Andrew out of my mind only during the Kol Nidre service. The issues still haunted me in the morning after a bad night's sleep. I was so upset that early Monday morning (before going to Yom Kippur service) I snuck into the office to meet with Andrew, telling him forcefully to keep out of the technology. From when the data comes into our algorithm through the SDC optimization was my responsibility. That was the only time in the 15 years I'd been affiliated with the synagogue that I had allowed work to intervene on Yom Kippur.

Andrew had formally been CEO barely three weeks, and I was tremendously upset. My window of latitude with UCLA had expired. For the annual UCLA report on Outside Professional Activities that comes up in October, I had been relieved to report that I held the board position (chairman) and a one-day-a-week consulting role – I didn't need to report an ongoing top-management job. Fall classes had begun, although my schedule was set up so that I didn't teach until January. This was to be the easy transition time so that everything was running smoothly and I could refocus on my UCLA efforts. Things were not going according to that plan.

I still had the power to get rid of Andrew. Four of the seven board seats were dedicated to the common shareholders, and I held the majority of common shares issued at that point. But this is nominal power, not real. As Penny often said, when the board of a startup begins having split votes, the company is doomed. While I feel this is somewhat of an exaggeration, it is largely true. In a board showdown, Andrew's vote would be frozen. It simply wouldn't matter if he voted himself in or out. Bud and Fred were the pivotal powers.

My discussions with Fred were what led to the quote given earlier that I should just lean back and let Andrew make me rich. At the $200-million valuation Fred had argued for a month earlier, I would have an eight-figure payday. Such a pot of gold is strange: It's not real enough to spend or even plan on spending, but it's too real to blow. At one point I asked Steve Mayer what $10 million meant. He said, "Think of it as an extra $50,000 per month, tax-free, forever." Oh. Fred's false premise, however, was that the technology was finished. When the screenplay is written, the director can take it from there. Our technology was still far too vulnerable to misapplication. The evidence I had so far indicated Andrew was clueless about solving the problems I knew we would face. Worse yet, Andrew couldn't accept that he was clueless. Fred indicated that the decision was mine. If we did get rid of Andrew, we would have to do it in a discreet, orderly fashion – perhaps hiding it for a few months so the business press didn't feast on the news to the detriment of our valuation. I didn't think the business press would notice very much, but Fred received more scrutiny. Forbes had just listed his membership on our board in its published summaries of the richest 400.

I inquired at UCLA about taking a reduced schedule in winter and spring quarters, in essence buying out course-time to focus on the business. It was too late to get a leave for the year; I would have needed to apply nine months earlier to comply with the newly established department policy. I heard back indirectly from the dean, through the department chairman, that the dean didn't want to create another Holtzman – an unkind reference to a colleague who maintained his professorship while his very successful outside company bought off half his courses. This raised the prospect of being forced to quit or retire. Retirement at age 56 was very costly in the UCLA system: The years between 55 and 60 had built in annual increases in the percentage credit for each year of service. Multiply the percentage credit times the number of years of service and the approximate retirement percentage results. I could retire at 56 or wait until at least 60. Those four years would increase my retirement income over 48%.

The issues with UCLA were far more complicated than the overt retirement percentages. I didn't start this venture as a substitute for UCLA. It was supposed to be a complement – providing a platform for the next generation of research in marketing science. Data and computer infrastructure drove developments in my field. Strategic Decision Corp. was supposed to provide both -- for me, and for the coming generation of academics such as Bradlow, Bronnenberg, and the others on the Academic Advisors Panel.

I talked with Bud about some of the practicalities. At this point he repeated his willingness to guarantee the next $5 million in funding. He balked, however, at adding the equivalent of my UCLA salary on top of my consulting contract, indicating that investors might think it too much for a startup.

If I left UCLA I was facing the certainty of short-term financial damage in total compensation and the certainty of long-term damage in UCLA retirement income. On the other hand, I had the long-term uncertain gains through a liquidity event with the company under my leadership, and some less likely gain through a liquidity event with the company under Andrew's leadership.

Does the entrepreneur choose to gamble for the big payoff? All the research in behavioral-decision theory indicates that people weigh certain loss much more than uncertain gain.[45] Entrepreneurs may be less risk-averse than others, but the mental algebra is fundamentally the same. What the savvy entrepreneur chooses is to do the homework that avoids this dilemma. I failed that test.

At a spring 2002 afternoon workshop called the UCLA Faculty Startup Seminar,[46] I put a question to the keynote speaker, Bill Sharpe. William F. Sharpe is the founder and chairman of Financial Engines, Inc., professor emeritus at Stanford, and the 1990 Nobel Laureate in Economics. He claimed to want to have nothing to do with the management of his company, Financial Engines. He wanted to turn that completely over to professional managers. I challenged

---

[45] See Tversky, Amos and Daniel Kahneman (1981), "The Framing of Decisions and the Psychology of Choice," *Science*, 211, 453-458, or Thaler, Richard.(1999), "Mental Accounting Matters," *Journal of Behavioral Decision Making*, 12, 183-206.

[46] "How to Start a Company without Quitting Your Day Job," March 20, 2002, UCLA Office of Research Administration.

that stance, saying that he was still a product manager responsible for the quality of the product that underlies his company. "What would you do," I asked, "if your CEO changed your equations?" Perhaps taken a little aback by the question, he responded, "I'd quit."

I didn't seriously ponder quitting. I had an exclusive consulting contract that ran until the end of 2001, and didn't want to consider the consequences of breaking it. Bud insisted that I stay involved to protect his investment and that of the investors he brought in. Other than Bud's assurance of funding, I didn't try to leverage his desire for me to stay involved into any other benefit. I chose to continue to try to work together with Andrew.

One of the many differences between Bill Sharpe's board and my board is that if he quit, I suspect his board would fire the CEO and beg Sharpe to reconsider. His board would take responsibility for solving the problem. If I quit, my board would insist that I solve the problem of how the company could possibly go forward without me. If I forced my board to fire Andrew, I would still have to solve the problem of how to continue without Andrew. So the point is not whether my board supported me. Rather, I had failed to get the board to accept its role in solving such thorny problems.

A subtler shift in board politics was also underway. Bryce approached Bud about investing in another venture in which he was involved. Steve sounded Bud out on a venture he was considering involving digital archiving and digital rights management. Penny and Wayne Levy, her partner in founding *efficient market services, inc.*, talked with Bud about yet another venture they had in mind. All of them checked with me on the propriety of these discussions. I had one criterion: if the proposed venture's success could help Bud's other ventures, I had no objection. But the cumulative impact was to make Bud even more powerful than he already was. To Bud and Fred, the technology was an invulnerable and impenetrable black box. I knew what was inside the black box, and I knew it wasn't invulnerable.

4.6   The Heart of Darkness

Once I shifted into a defensive posture, trying to protect the technology, Andrew started controlling all the external realities of the company. He took over the discussions with Goldman Sachs, which mysteriously changed from being enthusiastic about leading the next round to needing us to demonstrate another major victory in implementation before further funding.

Allowing this shift was a huge mistake. Fred had been yelling to turn our pilot success with iPlayer.com into immediate funding. While his $200 million valuation was outlandish, I thoroughly agreed with the need for funding based on the benchmarks just achieved, rather than some uncertain future implementation. Goldman Sachs had initially whispered an uncertain valuation between $75 and $150 million, but the new demands put time against us. Our successes so far, however, weren't Andrew's. I believe he wanted to have much more of his personal mark on our victories.

According to Andrew, our signed BrightStart.com deal at approximately $120K per year wasn't worth allocating personnel to implement. Once he saw that a Bizrate.com deal would start at half that, it wasn't worth doing. Building a forecasting and ordering product based on PersonalClerk for the home-video division of a major studio wasn't "strategic" to Andrew, regardless of the revenue it would bring in. The retrenchment at eHobbies.com and the widely heralded problems at eToys were the only justification Andrew needed for shifting all focus to the AdServer model. Everything except development of the AdServer model seemed to be put on hold.

The strong boundary I set up around the Office of Technology on Yom Kippur morning was under continual assault. Andrew's most organizationally detrimental retaliation came when I innocently replied to an email from Chas Frittoli (Carol's husband). Andrew called me on the carpet and read me the riot act for sending this email, which read:

> **From:** Lee Cooper
> **To:** Carol Frittoli, Jason Kapp, Andrew Harper, Chas Frittoli, Troy Noble
> **Date:** 10/16/00 5:52PM
> **Subject:** Re: (Anonymous ISP) Meeting (10/23)
>
> I think I should attend. We can determine who is best to present at a later time.
>
> Lee Cooper, Chairman
> Strategic Decision Corp.
>
> ************************************
> >>> Chas Frittoli <chas@xxxx.yyy> 10/16/00 04:20PM >>>
> Hello all,
>
> I have tentatively scheduled the follow-up meeting at (Anonymous ISP) for next
> Monday (10/23) at 11:00am. We will need to identify the team that will present our findings at this meeting. So, if anyone has a conflict please let me know.
>
> Thanks,
>
> Chas

To Andrew, sending this email was a gross organizational sin. After that I was forbidden from talking with people outside the Office of Technology except Andrew. All my communications were restricted to meetings where he was present. Otherwise I could speak only directly with him. Looking back, it was like sending a child to his room for being bad. At the time I resolved that if this was what he insisted on to keep his fingers out of the core technology, then I would comply.

The premise was wrong. Andrew didn't want to keep his fingers out of the technology. He was the CEO and the CEO can do anything he or she pleases as long as the board doesn't intervene. Andrew portrayed our differences as internecine squabbles – a founder who couldn't relinquish implementation to a new CEO, scientist versus professional manager. Except I was the guy who had trained professional managers for the past 30 years, and he was the guy with degree envy. Even if he didn't have the degree, he could be the boss of those who did.

Andrew started a management-by-threat campaign. "Explain the (Anonymous ISP) result or else I will cancel the NBCi test" -- despite the fact that he already had a top-line explanation of the (Anonymous ISP) test result[47] and the explanation was not on the critical path to a test with NBCi. This resulted in the infamous Friday night incident. Andrew was obsessed with demanding his explanation. One Friday night at 7 p.m. he was so upset I wasn't available that he gave a punitive all-weekend assignment to Ravi -- one of our most valuable employees and my direct report. The assignment was an ill-considered and naïve series of queries of a 270-million record file that required detailed SQL construction. When I received Ravi's message of this the next day, I cancelled the assignment and re-established the work schedule according to Andrew's own established priorities. To Andrew's credit, he apologized for this incident – admitting it was done because he was mad at me for not being at his beck and call at 7 p.m. on a Friday night. But that was done in private discussions. The email record of the events conveyed quite a different impression. Obviously very skilled at corporate infighting, Andrew's email traffic on this was so crafted that I shared it with Penny. Her impression was that Andrew was getting the better of the written record. That was when I started thinking of him as King Andrew the Specious. It was months before I started sharing that nickname with others.

Reflecting more on this, I believe that Andrew wanted to appear in these meetings with external agents, be they the free ISP, NBCi, Lycos, Goldman Sachs, or whoever, as if he knew all the answers. Our task was to fill him up with answers. Impossible. When I had been confronted with tough questions in due diligence, one of three things happened. First, I found ways of synthesizing elements from

---

[47] Most free or reduced-rate ISPs get revenue by enclosing their browsers in a frame that contains banner ads. These ads change every 30 seconds, regardless of whether that page being viewed is updated. 120 ads per hour rapidly depletes the ad inventory, meaning the same ad is repeated frequently even when all ads are shown. Showing only the best ads would wear them out even faster. We showed this in simulations based on their real data before ever moving to a live test with this ISP.

my background and stitching them into an answer on the spot. I wasn't interested in trying to fill Andrew up with enough background. Although I attempted some tutorials with Andrew, I resisted partly out of a sense that he hadn't prepared himself for advanced study, partly out of the prior reality that answering such questions was my role, and partly from finding that books I loaned him went unread. Second, I'd take wild intuitive leaps and speculate that we could do this or that. For the most part my intuition was correct, and the leaps were little more than long and loose stitches across pieces of knowledge. I watched Jason pale a bit as I indicated in early client meetings what future technology could do. It reminded me of many years ago presenting a paper at a prestigious invitational conference at Osaka University that my long-time Japanese co-author, Masao Nakanishi, had arranged for me to attend. At one point, my intuitive response to a question on the capabilities of our competitive market-share models caused Masao's face to turn completely white. He hunched over his note pad and, as I talked on, spent the next 15 minutes madly deriving equations. Finally, he put down his pencil, folded his hands over the note pad, and smiled at me. The color returned to his face. Masao may have saved my butt that day, but at least I was prepared. Andrew wanted to be given a PhD pill, take it with water and have all the answers he needed. And third when responding to tough questions I learned that when I started a wild leap, looked down, and saw nothing but an abyss, I could say, "I don't know. Let me think about that and get back to you."

Andrew's assault for control of the technology continued through the constant addition of features that he determined the client *must have*. Usually, product features are classified as *must haves*, *linear satisfiers*, *hygiene features*, and *delighters*.[48] *Linear satisfiers* are like price: the lower the better. *Hygiene features* are like clean bathrooms: Users aren't happy about a clean bathroom, but they are very dissatisfied if the bathrooms aren't clean. *Delighters* are those features that make a customer say "Wow!" The designs of the iMac, iPod, and PT Cruiser are examples of industrial designs that delight customers. *Must haves* are the minimal features a product *must* contain to be at all acceptable in the marketplace. Product-development experts warn of *feature creep* -- the slow additions of nice features that add precious time and costs

---

[48] Noriaki Kano, Shinichi Tsuji, Nobuhiko Seraku and Fumio Takahashi (1984), "Miryokuteki Hinshitsu To Atarimae Hinshitsu (Attractive Quality and Must-be Quality)," *Journal of Japanese Society for Quality Control*, 14, 2.

to development efforts. We had *feature gallop*, and I tried desperately but unsuccessfully to stop it. Every day new design documents were created and existing design docs were modified. Every time I tried to get a freeze on the product specification, Andrew would cancel it, and add more *must haves*. I again explained Christensen's Option 1 and Option 2, but Andrew wasn't interested in that lesson.

I met with Penny, Steve, and Bryce – more as my personal advisors rather than in their formal roles as the representatives of the common shareholders on the board. The sad truth I heard from them was that CEOs had the power to do what Andrew was doing. As long as the board doesn't fire him he can continue doing exactly what he wants. I had been the CEO for almost 11 months. I had thought my job was to communicate the vision, sell the concept to investors and potential employees, find the best people, figure out how to align their great skills with the tasks to be done, empower them, keep close watch on the progress they made, and help remove the barriers to progress. Now they tell me I could have just ordered them to do exactly what I wanted. Stupid me.

The seeds of crisis were sown when Andrew assigned quality assurance testing to Client Services, while I was restricted from speaking to the people within Client Services who were supposed to head up this effort. The Office of Technology created a testing suite called pTester, but it went untried for about three weeks. Carol Frittoli was initially put in charge of quality-assurance testing. Very able but already overloaded with other responsibilities in Client Services, she didn't try the testing suite immediately, and didn't find out until very late that she couldn't understand how to run the arcane command structure of the test program. She couldn't set up the conceptual business scenarios, etc. Without interdepartmental communication, these problems festered. We were in direct violation of the fundamental principles of cross-functional, product-development teams. Isolated in my silo, I couldn't manage this process.

Andrew's response was the epitome of one-sided hierarchical control:[49] "Come up with a testing plan or you'll all work Thanksgiving Day."

---

[49] For a thorough discussion of the traps created by one-sided hierarchical control see Culbert, Samuel A. and John B. Ullmen (2001), *Don't Kill the Bosses! Escaping the Hierarchy Trap*, San Francisco: Berrett-Koehler.

The rumors from Goldman Sachs were that valuation would come in between $50 and $100 million, depending on the results of an upcoming test with Lycos. Bud started talking about a $2 million bridge loan, rather than the $5 million he'd previously guaranteed. This was supposed to be the time we were regularly adding new clients to PersonalClerk. Instead we were grinding through development of a second major product. Our staffing was aligned with the needs in the go-to-market phase, with the higher burn rate associated with the anticipated client-support needs. Instead, the Client Services folks were tasked with product development, detailing specification that the overworked technology team had to program; and quality assurance, without the skill set to handle the required simulations.

Andrew had put the operation into stealth mode, throwing a Cloak of Invisibility over the whole company, supposedly to keep our competition unaware of our developments. This meant idling the Marketing Department and the associated trade-show staff, and keeping our advertising and public-relations agency, Young & Co., hanging on the line.

The time pressure was intense. While preparing for the Lycos test we were moving ahead with the implementation at iPlayer.com, which by then had been sold to another video-gaming site and was about to be sold to a foreign media conglomerate and integrated with its video-gaming network. The combination meant a shift from AdForce to DoubleClick's DART system for ad serving – a difficult transition. The addition of all these Internet game-playing sites increased the traffic expectations on our system, but the scalability of Giovanni's underlying architecture made this a competitive advantage for us. We heard reports of competitive recommendation systems that ran at 200 recs per second when we were hitting 1,100 recommendations per second in our most basic hardware configuration (i.e., a hardware load balancer and two VA Linux dual processors). The hardware was scheduled to be installed by December 1, followed by the installation and testing of the COM on iPlayer.com development servers by December 4, by which time we were also to complete the hook-up to DoubleClick's DART data feeds. Sales training and trafficking training were slated for December 4-8, along with complete quality assurance on the DART integration by December 6, full test integration on iPlayer.com's development servers by December 8, roll-out to the live iPlayer.com pages beginning by December 11, and full roll-out completed by December 15.

Still Andrew didn't come back with a signed contract from iPlayer.com, and wouldn't freeze the design specification. His efforts to lock Bud and Fred into a $2 million bridge loan (in which he would participate) were rejected. Fred, who two months before was arguing for a $200 million valuation, now was balking at a $27-million valuation, $2 million above the last round, even when the loan carried a 10% interest rate, and 25% warrant coverage at the Round C valuation. He now claimed this was very risky money, and he needed a better deal if he was to participate at all.

My approach to product design for the business constraints in the AdServer model was to set in place basic mechanisms. Plan A involved going back to learning who likes what ads when particular ads were not serving often enough to meet their budget goals (Daily Ad Caps) and turning off learning when no longer needed. Plan B involved allocating ads to unregistered users when Plan A still couldn't provide enough traffic to satisfy budgets. And Plan C was an almost mechanical set of gears and levers that could intervene on the normal preference priority in a segment-gender combination to move an ad higher if it needed more impressions or lower if it was over serving or under serving its budget. The traffic each day was estimated by a simple time series for each segment-gender cell. I knew some days would be overestimated and some days would be underestimated, but as long as the estimation wasn't biased, it would all average out over time. Having plans to control what we and let the rest fluctuate randomly fit my personality and the way I'd learned to model data over the decades. Of course, I underemphasized the role my personality played and thought of this as the *expert way* to approach the problem.

Andrew's approach couldn't tolerate leaving anything to chance. He threw out Plan A and substituted fixed budgets for learning. He trashed 90% of Plan B and concentrated on the gears and levers of Plan C. Every time a simulation would show a variance from the forecast, he would propose a mechanism to control it and change the product specification to correspond. Where I would set up 50% tolerances, he would cut them to 10%, thinking I was being irresponsible and the client would never accept such swings. I would forecast the traffic volume for a day, and he would want the system to forecast traffic every two hours. He micromanaged product design just as he micromanaged personnel.

Andrew couldn't stand my approach to product design and development any more than I could stand his. But he was the CEO and he simply drove right through me, as befits a football lineman, taking over direct control of all product development.

In an early-evening, mid-December meeting with Bud, Andrew, Bryce, and me, called to straighten out the differences between Andrew and me, Bud completely backed Andrew, indicating I wouldn't get funding without his presence. Bryce was largely silent. This meeting established Bud as being in total executive command of policy, and Andrew as the total operating authority. I was relegated to a newly created Office of Research that would consider long-range product issues.

On the positive side, I got to work with a new hire, Xuegao (pronounced shway-ga-o),[50] a UCLA PhD in nuclear engineering. You might wonder how nuclear engineering related to our efforts. Well, a nuclear meltdown, the *China Syndrome* as it was called in the movie, was the only thing less likely than a click on a banner ad. So nuclear engineers were well versed in the statistics of the extreme processes. Further, he understood the mathematics I used to characterize probabilistic processes. Even Murilo, the PhD physicist and fine mathematician who programmed the business-rules module, wasn't familiar with statistical computing. He had to translate my math into some mental algorithm that he could program, and some things got lost in translation.

On the negative side, I felt like I had deserted the great people in the Office of Technology to a Philistine. Murilo was caught in a bind. Simulations were run overnight every night, from scenarios that became increasingly bizarre as Andrew tried to exert micro control over a messy world. In the morning the Client Services staff assigned to product management would lurk over Murilo as he reviewed the output for possible code errors or other variances from Andrew's expectations. His report became the fodder for a staff meeting involving Andrew and key personnel in technology and Client

---

[50] Ravi's expertise in Chinese names taught us what UCLA never learned. Xuegao had been know as David at UCLA due to Xuegao's desire not to have to teach everyone he met.

Services. Murilo faced an oral exam on the detail of each sub-routine. Andrew would arm-wave his way through an alteration. Ravi and Giovanni would have to figure out how to translate arm-waving and bullet points into code. And Murilo would have to code it and integrate the changes into the bulging structure of the business-rule algorithm. Giovanni started to mumble about the *spaghetti code* that resulted.

Throughout, Fabrizio remained calm, due more to his serious martial-arts training than anything else. He could focus on his needed actions regardless of the clamor. Giuseppe was never calm, but was unchanged by most of these events. Wesley was frozen by any interaction with Andrew, and thus was eventually excluded from these meetings, though he continued to do excellent database work. Jonathan chafed under Andrew's tight-fitting collar. In the middle of a rather intense meeting, Jonathan blurted out, "That's actually a good idea," when Andrew for once said something that made sense to him. The room was overcome with stifled, embarrassed laughter. Ravi was the unflappable gentleman, contributing his great intellect on both the technical and managerial side. Giovanni was ever the maestro -- tirelessly re-orchestrating the efforts after Andrew ripped pages out and pasted them into the score. I apologized to each of them for my loss of control over their work lives. I'm sure this undercut Andrew's standing. Perhaps that is what I meant to do.

In 10 days Xuegao and I developed a much simpler and more robust alternative to the spaghetti code. Based on standard choice-model principles, we called it the Share-of-Choice (SOC) method. For three months, no amount of positive results from our testing or simulation convinced Andrew to implement it. He always found some excuse not to believe the test results. This might be attributable to solid business conservatism if it were not for counter-examples such as what occurred on Martin Luther King Day in 2001.

After a 9 a.m. meeting to set parameters for Lycos execution, Andrew came into Giovanni's room where Ravi, Giovanni, and I were talking. He said he wanted to try cutting out of execution all cells in which an ad's click rate was below average for that ad. This represented an extremely major conceptual change and substantial coding effort to be implemented and tested within 10 hours of beginning execution at Lycos. No design document (the standard procedure for any candidate for implementation) had been developed for this approach.

No discussion of this approach had been aired at any design meeting. For those familiar with this technology, Andrew's premise confuses between-cell variation in click rate with within-cell variation in click rate. It could lead to there being no ad served to people in certain segments. Other complexities had never been thought through. I explained this to Andrew, to no effect.

I told Andrew that I thought this was a tremendously undisciplined and irresponsible act. I went to my office, called him and asked him to come talk to me. He said he would in 30 minutes. During that time he set in motion the code changes to implement his scheme. He planned one test scenario, and if it improved click though rate, he would put it into execution without further testing. In my office shortly thereafter, I repeated my accusations, telling him that he was betting the company on an unproven concept and untried code.

He didn't implement this code because it did not improve click rates in the one simulation he ran. What if it had been marginally better in this one test scenario?

After months of unsuccessfully trying to get Andrew to implement the Share-of-Choice approach, Xuegao quit. Talented people don't stick around if they sense their talents are being wasted. The interaction on this solidified the reputation of *King Andrew the Specious*. He would look for anything good in the ideas he favored and anything bad in those he didn't. That aspect became the most important thing —a stopper if he opposed the approach or a reason to overcome all other obstacles if he supported it.

4.7   All Work Is Voluntary

At one point Giovanni looked at me sadly and said, "What can I do? Andrew is my boss." I replied, "I was your boss for four years. The difference is when I was your boss you never knew it."

The maxim one always hears is, "*A*-people hire *A*-people, while *B*-people hire *C*-people." *Human risk* is one of John Doerr's big-four risks: Can the venture gather the human capital to master both the technology and the management needed to bring it to market? Brian Farrell, the chairman, president and CEO of THQ, Inc., is just the most recent keynoter to emphasize in his 2002 address to the UCLA Entrepreneur Association conference that you must hire people smarter than you are.[51] These platitudes beg two questions: Why bother? And, Why would *A*-people want to work for you?

The first question is easy. Entrepreneurial enterprise extends beyond the capabilities of any one individual. While replication extends enterprise in *scale* – more B-people or C-people can extend the scale of routine work – replication will not extend the *scope* of enterprise. The broad scope that characterizes entrepreneurial vision requires confronting unknown territory. *A*-people possess the problem-solving and learning skills that translate a vision in an unfamiliar context. A top programmer is 10 times as productive as an average coder, not because he or she types 10 times as fast, but because of the conceptualizing skills that lead to a simpler solution to a programming problem.[52]

The second question is subtler. It is tempting to assert that you get top talent by paying top dollar. But the truth is that labor markets for top people are competitive. The staff at Strategic Decision Corp. was paid well, but the times demanded that. We never provided a lunchtime massage that some startups offered. The refrigerator was stocked with sodas and snacks as much to minimize time lost going to the snack bar on the ground floor as to provide a more richly appointed work environment. The top-notch espresso maker was the only conspicuous luxury, and Giovanni and Van paid for that themselves.

Anywhere *A*-people looked they would have their basic needs met: physiological (i.e., hunger, thirst, and bodily comforts), and safety and security (i.e., staying out of danger). *A*-people expect that their needs for belongingness (i.e., affiliate with others and be accepted) and esteem (i.e., to achieve, be competent, and gain approval and recognition) will be met in their work life. *A*-people also have a cognitive need (i.e., to know, to understand, and to explore).[53] They *choose* to go where these needs can be met. Because the basic needs are taken care of, and because of the choiceful-ness involved, I believe it is essential to think of all work by top people as *voluntary*.

---

[51]Morning Keynote, EA Conference 2002 "(re)Building Blocks for Entrepreneurship," May 12, 2002.
[52]Brooks, Frederick P. Jr. (1995), *The Mythical Man-Month: Essays on Software Engineering Anniversary Edition*. Reading Mass: Addison Wesley Longman, Inc.
[53]Maslow, Abraham (1943), "A Theory of Human Motivation," *Psychological Review*, 50, 370-396. Maslow, Abraham (1954), *Motivation and Personality*, New York: Harper.

So, if you are lucky enough to be in the position of recruiting truly top people, the hiring decision becomes selecting *who* among the top candidates will *voluntarily* go the extra mile that startup companies require. Look for the volunteers. The managerial challenge becomes how to present the problem terrain so that your people *voluntarily* take needed responsibility. The implicit exchange goes something like this: You believed deeply enough in this enterprise to voluntarily commit yourself to this effort. Your valued co-workers are voluntarily taking their responsibilities. Of course you will volunteer for your part and take the responsibility to see it through. Thinking of *A*-people as volunteers doesn't lead to pampering big egos and loss of managerial control – almost the opposite is true.

The entrepreneurial vision charges the mission of the organization with value, much like love for the arts charges arts organizations with value. This process is akin to charging a battery (i.e., filling it with potential energy) or establishing a bank account with psychic income. This value can be used in marketplace exchange – at times more tangibly than dollars. And this value attracts voluntary participation by *A*-list employees. Angel investors and venture capitalists exchange hard dollars for a piece of this value.

The clarity of the direction articulated by the entrepreneurial vision helps control organizational behavior. *A*-people internalize that vision, understand where the company is going, and because they have voluntarily chosen this affiliation, shape their behavior to advance organizational goals. This is the organizational-behavior version of what in social psychology and marketing is called the foot-in-the-door technique, where an initial, perhaps small, voluntary commitment legitimizes much more major voluntary commitments later on.[54] Key to this is igniting that spark of volunteerism.

We hired great people with this implicit approach. They willingly put in long hours and great, creative effort in pursuit of common goals. And I allowed an autocrat to come in and treat them like wage slaves.

---

[54]Cialdini, Robert B. and David Schroeder (1976), "Increasing Compliance by Legitimizing Paltry Contributions: When Even A Penny Helps," *Journal of Personality and Social Psychology*, 34:599-604. Cialdini, Robert B., John T. Cacioppo, R. Bassett and J. Miller (1978), "Low-Ball Procedure for Producing Compliance: Commitment Then Cost," *Journal of Personality and Social Psychology*, 36:463-476. Scott, Carol A. and Richard F. Yalch (1980), "Consumer Response to Initial Product Trial: A Bayesian Analysis," *Journal of Consumer Research*, 7 (June), 32-41.

I am profoundly embarrassed by this, and still kick myself for allowing it to happen.

Why didn't they quit when the regime changed? Some did. We lost all of our PhDs – one before Andrew took firm control and four after. Giovanni finally completed his dissertation in the spring of 2002, so the company now again has a PhD on the payroll. Some stayed partly because they were new parents. A burst of procreativity accompanied our new-venture creation. The VanArsdales had a daughter, as did the Kapps, the Pennington-Penders, the Yus, and the Srinivasans. The Giuffridas, DuPont-Bronnenbergs, and Hendersons had sons.

Most of the others we lost through layoffs. The first wave of layoffs came at the end of January 2001. As I indicated above, our staffing was set for rapid market deployment of our e-commerce product, PersonalClerk. When Andrew allowed development of the AdServer model to drag on and on, our burn rate was unsustainable. The whole Marketing Department was laid off in this first wave. I felt I should be the one to tell Kate, the VP of marketing who was there from the beginning. We had a very emotional session, not over the job loss – Kate was very able to find alternatives -- but over the loss of the storybook fantasy of the company. We agreed on the vision, used our ideals to build something together, and now it was being dismantled.

The 37 full-time employees in January 2001 dropped to 27 by month's end. We cut all four in Marketing, three from the eight in Product Management/Client Services, two of the 12 full-time technology team (plus two part-time programmers who were laid off), one of the research staff that only had three including me, and one from the Admin/HR staff of six. The sales team of two and the top executive team outside marketing stayed intact. We held an all-hands meeting the morning after the layoffs. Andrew asked if I wanted to speak. I told the survivors that we shouldn't forget the people who were gone. We shouldn't act as if the layoffs hadn't happened. I said the people who were gone worked as hard and contributed as much as those who were still here. They were gone because the path to market was different than we planned; to remain viable as a company we had to cut our burn rate to the essential minimum needed to complete development and implementation of our product.

I spoke first, since I had office hours and class later that day at UCLA. I didn't stay to hear Andrew's remarks.

Any academic turned entrepreneur who, after reading this, still fails to do the required investigation of a potential CEO gets what he or she deserves. The company being created, however, deserves better. When I interacted with potential CEO candidates I was often in *sell* mode, trying to convince a seemingly desirable candidate to take on the challenge. Asking probing questions seemed out of place. The time is so pressured that looking into old S-1s or 10Ks seems like not a good use of so precious a resource. The word of trusted advisors and the surface record of accomplishments are not enough. Get the facts and ask the questions as if you were giving your baby up for adoption. Charles Ferguson was lucky. It only cost him money.

5. Smart Money

This chapter explores the $1.25 million Series-A funding and the $5 million Series-B funding: how these deals were structured, how the parties positioned themselves to get a better deal than they might otherwise have obtained, and how all of us were influenced by the seemingly infinite availability of venture capital. I then report on the very altered funding environment the company faced in the later rounds.

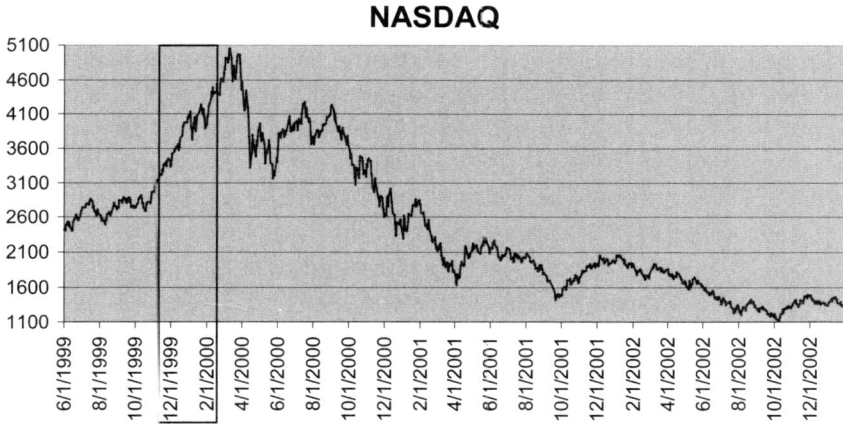

5.1 The Series-A Negotiations

What started as an informal exchange between Bud and me over lunch at the 17th Street Café on December 23rd 1999, had to be translated into a legal agreement. About the only thing that remained relatively constant was the valuation: $1.25 million for 25% of the company equated to a $3.75 million pre-money valuation and a $5 million post-money valuation.

The first negotiation was between Giovanni and me. We both knew that this wasn't the consulting partnership we had talked about years before. I brought the prestigious board of directors and advisory panel, the knowledge of technology-enabled marketing and marketing-science models, the results of my re-analysis of the 1990 U.S. Census that became ZipSegments, and the connection through Fisher Pennington to the expected clients and investors. Giovanni brought the KDS and NOAH algorithms and his computer-science savvy. To Giovanni, contributing these algorithms loomed as a big sacrifice. He thought any alternative future company he might want to create or independent consulting he might do, if this company didn't work out, would be compromised by assigning the rights to these algorithms to SDC. I knew KDS and NOAH would help establish our credibility and fundability, but guessed they would not find direct application in operating code. These algorithms represented tangible exemplars of what we could do, not what we *would* do. Yet, I could not say that directly to Giovanni, partly because I could not communicate that speculation with investors. We agreed that my share would be twice his and would be in common shares already vested. I felt my having shares rather than options reflected both the starting contribution and the need to maintain voting control of the company. Half of his shares would be already vested common shares, and the other half would be options that vested completely in two years – a one-year cliff for the first half of the options, followed by monthly vesting until complete. This whole negotiation was harder to conclude than you might expect. Giovanni wouldn't agree until he could *feel good* about the deal – leaving behind the notion of an equal partnership.

Fisher Pennington and Lennon & Ortega were to split 5% of the company that would not be diluted by up to $5 million in funding. That meant when the company had received $5 million in funding, the law firms would own 5%. To insure this, some of the common shares from Giovanni and me had to be set aside into a speculative share pool that would go to the law firms when the valuation of the company in the subsequent round of investment was determined. The set-aside share pool was done at the Series-A price. If the Series-B share price were higher, we would get some of that reserve pool back. When the Series-B (post-money) valuation was established at $25 million, this interest translated into a hypothetical $625,000 interest for each firm. I had no doubt that the strategic value of the Fisher Pennington connections would be worth that and more, though it seemed a little excessive for what I expected from Lennon & Ortega. But Fisher Pennington set up that deal. Without Lennon & Ortega, the whole 5% would have gone to Fisher Pennington. We'll return to this a little later in this chapter.

The Series-A investors wanted their 25% of the company to be as fully diluted as possible. That is, their shares would be X where $X/Y = 25\%$. They wanted Y to include as much as possible of the future share needs that the company could anticipate, thus making X larger and saving the Series-A investors from later dilution. The compromise we agreed on included options for Giovanni, Van, Jason, Kate, and Troy, plus options for the three board members representing common shareholders (Steve, Penny, and Bryce) that were added to Y. We arbitrarily determined that 10 million shares would reflect common shares already issued, and added 3.2 million common shares in reserve for the options mentioned above. That meant 4.4 million Series-A preferred reflected 25% of the company -- $.284 per share. In addition, the Series-A investors agreed to add 3,863,415 shares for future employee options. The value of the common shares was set at $.07, about one-fourth the value of the preferred. Thus, when the strike price for the options was set at $.07, no taxable event occurred. Of course, this also means that until a public market exists for all the stock, the common shares are worth much less than the preferred.

The other terms in the Series-A agreement details concerned the preferred return, liquidation preference, anti-dilution protection, board representation, compensation committee, actions requiring approval of the board, Series-A investors' reserved voting rights, information rights, restrictions on transfer, registration rights, and lock-up period.[55]

The founders (Giovanni, the law firms Fisher Pennington and Lennon & Ortega, and I) were represented in the negotiations by Edward Lennon. Alton Clark of Wilson Sonsini Goodrich & Rosati PC represented the Series-A investors. Conflicts of interest abounded since Bud, Paul, and many other attorneys at Fisher Pennington were investors. Edward Lennon was supposed not to invest in this round because of the conflict of interest, but when the negotiations were essentially complete, Bud OK'd his participation as an investor.

I didn't receive the conflict-of-interest waiver from Lennon & Ortega until 36 hours before the closing. The deal reflected the good market conditions of the time (January 2000), so I wasn't concerned. The

---

[55] An explanation of these terms appears on the VentureDevelopmentProject.com Website.

terms of engagement with Lennon & Ortega were included with the waivers and were very different from what I expected! As I indicated in Chapter 3, my understanding was that for its 2.5%, Lennon & Ortega would do the long-form work to create the company, develop a financial structure and employee contracts, etc. Once the company was structured and financed, we would receive a 25% discount on subsequent long-form work. I also understood that Lennon & Ortega would orchestrate the road show for the next round of funding. The document I received said something quite different. According to Edward Lennon, the 2.5% was for "(i) the 25% discount in hourly fees ... and (ii) the fact that I and others at my firm have performed unique advisory services for the group for which the group has not incurred any hourly fees, including, without limitation, by crafting and refining SDG's mission statement and business plan and creating and finding financing sources and other business opportunities for the group."[56] Once I had gathered the early management team of Jason, Van, Kate, and Troy, I had found his efforts totally superfluous. I had tolerated Edward's suggestions, because I thought it made him feel good. Now I discovered that these were the "unique advisory services" that were worth so much. I found out even later that hours for the long-form work establishing the company were not part of the deal, just the 25% discount applied. In discussing my different understanding with Edward, I asked why I hadn't been billed previously, so that I could have discovered this discrepancy before it was so late. He replied that the bill for the hours was carried as a loan until we had funds. I couldn't imagine that he had explained that to me and I just forgot. I still can't. I tried to check my recall with Paul Brendl, who was at the October 1999 meeting. Paul came up with the neatest dodge I ever got from him. He said that he heard Edward discuss related terms in so many different contexts that he couldn't say for sure what was said on that particular occasion.

It gets worse. The engagement letter informed me that Lennon & Ortega would ship out much of our work and receive a kickback from the referral of up to 20% of the fees billed to the other firms. Specifically, Riordan & McKinzie would be used extensively in this closing and had agreed to a 10% kickback. The totally legal arrangement meant that rather than providing a 25% discount on long-form work, SDC would be billed by a subcontractor at its full

---

[56]Letter from Edward Lennon dated January 11, 2000, detailing terms of engagement and conflicts of interest.

rates, out of which the sub would kick back 10% to 20% to Lennon & Ortega.

Regardless of what said in the original meeting, any attorney whom I so readily misunderstand is the wrong attorney for me. For efforts I thought was part of the 2.5% agreement, we ended up paying more than $65,000 in fees to Lennon & Ortega for work it had already done and committed to doing before I could extricate SDC from the relationship in early April. No formal termination occurred; we just stopped sending them work and inviting them to board meetings. After that, we worked directly with Gary Montoya, the Riordan & McKinzie partner who actually organized the Series-A closing. He made sure the work got done without a middleman. Whether Riordan & McKinzie continued to pay a kickback to Lennon & Ortega, I do not know.

5.2   The Prelude to Series B

The push for the Series-B funding began in earnest on March 1. Bud gathered a group at his office to hear the latest version of the pitch for a $2-million round that would take us to the fall. Reza Mamoun, who ran a private-client group at Merrill Lynch, came and brought Fred Hatchet, the CEO of iPlayer.com, and Andrew Harper – the first time I met Andrew. Reza arranged for Joel Wilde, a PhD and CFA from Intelligent Technology Ventures to come, not as a potential investor but rather as someone knowledgeable in this arena who could help ask the tough questions. Someone from the audit side of Deloitte & Touche, came representing a senior partner.

The former chairman and CEO of two major movie studios also attended. Bud had introduced him to me merely as Alan, the friend who convinced Bud to build a place near his own in Jackson Hole. I sat next to him by happenstance rather than design. As Jason prepared and delivered the presentation, he and I had a quiet, running Q & A. He asked simple, good questions with an avuncular curiosity that made me feel very comfortable. I liked him immediately.

Joel Wilde, whose fund had invested in a startup, CustomLink, involved in Website customization, asked pointed questions that tried to cast SDC as a CustomLink clone, at least in terms of the corporate-development path, if not the technology. Jason handled most of his queries. The email exchange that began the evening after the meeting captured much of the flavor.

> From: jWilde@xxxx.yyy [mailto:jWilde@xxxx.yyy]
> Sent: Wednesday, March 01, 2000 10:51 PM
> To: Lee Cooper
> Subject: Re: Strategic Decision Corp presentation
>
> Lee,
>
> Thanks for the info on Strategic Decision Corp.
>
> I like the space you are playing in, however, we feel very strongly that your financing strategy of raising small amounts of capital every few months at very high valuations from unsophisticated investors (i.e., dumb money) will in the end likely be successful at minimizing the dilution of the founders. However, very few, if any, college professors have ever gotten rich by owning a huge percentage of nothing.
>
> Your management's recurring focus on raising capital will weaken SDC from an operating perspective, and distract management from its central mission, to create a sustainable and fast growing Internet business where time to market, rather than "dilution of owners' equity" is the most critical performance measure. In addition, your piecemeal financing strategy could backfire severely, leaving the company insolvent if a major downturn in equity markets causes angel money to dry up and disappear.
>
> As an institutional fund, we have a fiduciary responsibility to our shareholders not to over-pay for companies in our portfolio. We would normally target a minimum return of 10 X our money to compensate for the significant uncertainty regarding the performance of your technology in a real trial. Given that CustomLink just sold for $190 million, there is a significant opportunity for the taking.

I replied:

> Lee Cooper <lee.cooper@xxxx.yyy> wrote:

Joel,

I greatly appreciate your comments. I am more interested in building a successful company than in saving the founders from dilution. I'd like a little more feedback on my current strategy.

I know that *smart money* comes to the table with more than money. We hope the VC we bring in will have the fat Rolodexes (Palm Pilots?) that tap a network of potential clients and allies. To me it seems best to get them involved at the point where we have a proven product (three referenceable accounts) and the staff in place to meet the scale of demand generated by those contacts. Right now we have the prototype, the alpha version will be available through a Web interface on Monday, we last night got the thumbs up from eHobbies for a beta test, and Friday I expect the OK on another beta that will most likely go live with our technology before eHobbies. We have a great set of VPs and CTO and a total staff of 13. In two months we should be ready to respond to increased demand from VC contacts. Doesn't it make sense to fund that gap with private placements?

Valuation. When I presented this idea first in October, I had two algorithms with three years of development behind them and a plan for solving the business design issues that could make a business out of what otherwise might be a consultancy. In particular, I knew how to cluster the U.S. Census to get the actionable triggers for our segmentation scheme (now called ZipSegments). A small group took the leap of faith required to build a business around this. For their participation and guidance I was very willing to accept their pre-money valuation of $3.75 million. But everything we promised in October is here on schedule in March. What is the proper valuation for that? Do you have a sense of what the valuation would be with the three referenceable accounts in the next step?

Any other comments on the timing of VC participation would also be appreciated.

Thanks again,

Cheers,
Lee

He wrote back the next morning:

> Hi Lee,
>
> 1) You need to do a more thorough competitive analysis, and clearly show in a product attribute matrix the key differentiators for SDC versus other players in the space, e.g., Epiphany, Blue Martini, CustomLink, Net Perceptions, Personify, etc.
>
> 2) You are an alpha/beta stage company in hot space. At this point in its development, CustomLink had a pre-money valuation of $11 million. I think $25 million in Southern California is definitely aggressive.
>
> 3) Obtaining three reference accounts may take longer than you think, particularly if the customer must test the productivity of your system in up-sale, cross-sale, and customer-service applications.
>
> If there is flexibility on valuation I would like to schedule a meeting with my partners.
>
> Thanks,
>
> Joel

The *valuation* played such a prominent role in this exchange that I felt I should get Ed Pinter's advice on how to deal with that issue. I sent him the email thread. He replied that afternoon:

> I assume (and think that I recall) that Wilde is a venture capitalist. Assuming that my assumption is correct, I'm leery of signaling too much flexibility on valuation. I fear that word will get out that the valuation will be lower and you'll never get it back up this round. I suggest responding to his points as follows:
>
> 1. Give him a succinct explanation of why SDC is different than all the competitors. For example, "SDC is the only company that can make recommendations in a flexible, efficient and user-friendly way without any privacy issues. Most of the competitors (e.g., Engage) are just selling advertising by detecting browsing

> patterns. Others make recommendations (e.g., Net Perceptions), but do so through a collaborative filter that finds statistical correlations among products without regard to the customer. Such correlations produce cruder, less useful results." I know that's not complete or maybe even fully accurate, but it's the tone that I think you need.
>
> 2. On valuation, I suggest something short like, "Rather than debate valuation, why don't we sit down and go over the product and its attributes. Afterwards, we can argue about valuation."
>
> 3. I agree that you should indicate that you're going for more money this round. I'd give him some of the credit for your decision in this regard (e.g., "As you suggested, ...").
>
> 4. I think that you need to address his point about the difficulties of obtaining reference accounts (which I assume are beta accounts).
>
> Ed

With Ed's input I fashioned the following reply:

> Lee Cooper <lee.cooper@xxxx.yyy> wrote:
> Hi Joel,
> Your point about "living on the edge" struck home with us. The time devoted to our anticipated two small rounds of financing will detract from important product-development efforts. We have decided to revert to an earlier plan and seek $4-5 million in this round. We would like to meet with you and your partners to discuss our efforts further and learn what else you bring to the table as a potential venture partner.
>
> We will do a more thorough competitive matrix. With the rapid evolution of this market we scrapped the big matrix in favor of one that summarizes classes. But I clearly need the complete version to handle queries such as yours concerning CustomLink.
>
> One feature that distinguishes us from any other play is the network externalities we create by combining a standardized segmentation scheme with our recommendation engine. This creates a segment language that retailers can use to communicate

> cross-sell opportunities between them, while completely respecting the privacy of each other's individual customers. Retailer X can arrange to cross sell the most popular "Segment 12" offer to "Segment 12" customers on Retailer Y's site.
>
> You may be right about the length of time to get three referenceable accounts, although in the last two days we received OKs on two beta sites (eHobbies and eAssist).
>
> Rather than debate valuation, why don't we sit down with you and your partners and go over the product, its attributes, and how you can contribute. Afterwards, we can argue about valuation.
>
> Thanks again for your feedback.
>
> Cheers,
> Lee

Apparently that was the right approach. Joel indicated that he would put me in contact with Joe Parks, a JD-MBA from UCLA who worked on the CustomLink deal, and Joe would set up a meeting. Jason and I pitched Joe on Wednesday at Intelligent Technology Ventures' Beverly Hills offices. That was followed shortly by a pitch at our offices to David Nalbanian, the principal funder of Intelligent Technology Ventures. Joe Parks followed that by spending almost two days at our offices, interviewing every person on staff. On Friday afternoon Giovanni took Joe to the nearby Indian restaurant. The spicy food, washed down with more wine than Joe normally consumes at lunch, made for a jovial final afternoon of interviews. Joe was fond of counting lines of code as a measure of development progress. I guess the wine and our numbers impressed him. Nalbanian wanted $3-5 million of the next round, which was only scheduled for $4-5 million. Remember, this is the middle of March 2000. The NASDAQ closed at a record high of 5039 on March 10, 2000.

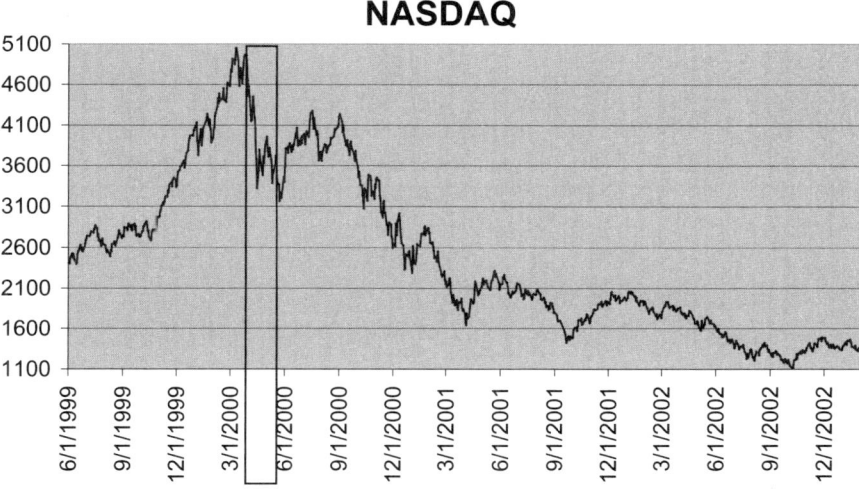

We had also been working with AT Kearney. The head of the West Coast office was so impressed that a startup could lure Jason Kapp away that he wanted to fund a strategic alliance. After a series of due-diligence hurdles, AT Kearney offered to provide $1 million in the B Round, 55% of which would be allocated to consulting services that AT Kearney would provide SDC. These equity-for-services deals can work out well for all parties, if the services are central to building the company. At this point, however, no specific services were planned. It didn't make sense for SDC to dedicate more than $500,000 to unspecified services. We asked that the investment be in cash. When this didn't fit AT Kearney's policy, we put AT Kearney at the bottom of the potential-funder list.

Bud, Fred, and Reza led the three investor groups that we considered in a March 23 conference call. Including AT Kearney, pre-emptive rights, and the $3-5 million that Intelligent Technology Ventures wanted, $10-12 million was on the table. Obviously, in hindsight, we should have taken all the money except AT Kearney's. The plan I had developed projected a positive cash flow before the $5 million would run out. The very savvy advisors on the conference call were also investors in the A Round (except for Reza Mamoun). That spelled conflict of interest. My interests as a shareholder aligned with theirs. We all wanted to minimize dilution, while taking enough capital to execute our plan. Here, however, the interests of the company were not aligned. The rule of thumb is that it always takes twice as much money and time as you expect. That cushion was already factored into our plan. Van had created an alternative scenario that would stretch the capital to cover the company for 22 months. What I had not put in my plan was the new-CEO factor. For Van and me to cut back, if circumstances warranted, was one thing. It was quite another to run into rough water when a new CEO was trying to establish control. The turbulence of the environment and the turbulence of the change could, and did, commingle in confusing ways.

The NASDAQ had dropped to 4455 by March 16, and risen back to 4865 by the 23rd. The end was not in sight -- because we were looking ahead. The high had already passed us, while we were too wrapped up in the swirl to see it.

I took AT Kearney's offer off the table. My discussions with Intelligent Technology Ventures indicated that I could keep them in as long as their allocation was around $1 million. In addition to Intelligent Technology Ventures, the B Round included two New York VC funds: a media-oriented fund sponsored by Bear Stearns, and INET Ventures (a publicly traded venture-capital fund created by the former studio chief I mentioned earlier). With these venture funds, the issue was making sure they had enough skin in the game to open their Rolodexes for us. Bud indicated that he could convince all his partners not to exercise their pre-emptive rights. Bud and Fred also agreed not to exercise their pre-emptive rights or put new money in on this round. The argument was that this round should include the new money that came with the connections the company needed to succeed. This is the *smart* in smart money. It has nothing necessarily to do with savvy financial investors. The media fund sponsored by Bear Stearns connected us with email-behemoth Bigfoot and database-gorilla Acxiom. INET could connect us with CBS MarketWatch. Intelligent Technology Ventures connected us with Internet retailers Union-Shopper.com and BrightStart.com. Andrew Harper connected us with iPlayer.com. Other individuals in the B Round connected us with Goldman Sachs, Merrill Lynch, National Securities, and the Yucaipa Company.

Given all the demand, I pushed for a high valuation. The valuation number was really going to be the largest number Bud could communicate to his investor group without feeling uneasy. That number, pre-money, was $20 million. Done. Despite the earlier posturing of Intelligent Technology Ventures and a few other major check writers, they all wanted in when they heard the final valuation.

Everyone would have to take a haircut off of what they wanted to invest, but without AT Kearney and pre-emptive rights, we could accommodate the demand. We had a hard time getting a check from one billionaire whom Fred brought in. Because of his late statement of interest, he was allocated only $200,000. For that little, the request for a check kept falling to the bottom of his in-basket.

Alton Clark from Wilson Sonsini represented the B investors and Gary Montoya from Riordan & McKinzie represented the company and the Series-A investors. Alton needed to know whom he should work with among the Series-B investors. He worked with Fred and Paul in the Series A, but neither was making an investment in this round. Alton suggested Andrew Harper. Although he wasn't the largest investor in the B Round, more had gone to Intelligent Technology Ventures and INET Ventures. This seemed fine to me, since his name was already in play as the B-Round representative on the board. We agreed, basically, to update the terms in the A Round. The company was more mature and better financed, so spending limits were increased to $250,000 prior to requiring board approval. Reza Mamoun was adamant about including 13% of the shares in options for future key employees (CEO, CFO, VP Sales, and others). Protecting 5% for the future CEO was high on his agenda, I'm sure. I always favored the employee options, but wanted the Series-B investors to share in the dilution. Mamoun insisted that the Series-A and common shareholders shoulder the dilution themselves. This dropped the share price from $1.388 to $.952, boosting the number of shares we had to issue to the Series-B investors to 5.25 million. These details were concluded in a pressure-packed phone call between Reza and me on April 14. As the NASDAQ fell to 3266 that day, Reza said if I didn't like his plan I could always renegotiate the deal with the investors. Amid that kind of turbulence, renegotiation seemed very unwise.

I wanted the B-Round documents to set the precedent that one board seat should go with each major investment cycle. Boards that grow large are too unwieldy. Even setting agreeable meeting times becomes impossible with too many very busy people. I asked for and got veto authority on one suggested board representative on this round, and hopefully future rounds. That means if the investor group suggested Attila the Hun be the board representative, I could veto him. If the second suggestion was worse, I was stuck.

5.3 Don't Even Think About a Down Round

While Chapter 7 looks more carefully at the collapse of the Internet advertising market, here I'll focus on the narrower implications for financing Strategic Decision Corp.

The test with Lycos was a disappointment. We achieved about 2× lift, but had expected much more. The part of the site on which we were allowed to test had ads that changed every 30 seconds regardless of user input. This was exactly the situation that the simulations with data from the free Internet services told us to avoid. Inventory was worn out regardless of our best efforts. I was critical of Andrew and Jason for missing such an obvious problem. They were critical of me for not reminding them of the obvious. This is the kind of problem you can expect when the person who understands the marketing models is kept away from the client. The test with NBCi was cancelled due to that company's lack of appropriate customer data.

Product development struggled on with iPlayer.com, as mergers and acquisitions precipitated changes in corporate identities. Traditionally, the big money was associated with the cost-per-thousand (CPM) side of this business, but CPM rates declined and the diversity in advertising inventory shrank. We signed revenue deals for the cost-per-click (CPC) side, but while we continue to develop for the CPM side, no revenue-sharing agreement was forthcoming. Remember that increases in click rates translate directly into revenue on the CPC side of the ad business. On the CPM side, the effects are indirect. Higher click rates should make it easier to sell CPM deals, but when a site has excess inventory (i.e., more eyeballs than paid ads to show), finding a market-clearing price for a CPM deal is difficult. Should the site take a CPM deal at $1 CPM? Only by translating the click rate on a CPC deal into the equivalent CPM rate can a rational floor for CPM prices be established. If the existing business patterns arbitrarily limit the number of CPC deals, then even this rational pricing mechanism is clouded.

Approaching the March 2 board meeting, bridge financing was still not in place. We needed $1.5 to $2 million, and Fred had signaled Andrew and Bud at the prior valuation that he wasn't willing to put up his part of the bridge loan. The NASDAQ was down to 2100 and Ed speculated that most of Fred's investments were under water.

Ed replaced Bryce on the board of directors. Board terms were year-to-year, but Bryce expected to stay on, since Penny and Steve were staying. In the breakfast meeting where I told Bryce of this, he said he felt I was punishing him for not speaking up for me in that harsh mid-December meeting with Bud and Andrew. I denied that, and in truth it was at most a minor part of the reason. I certainly didn't take his silence personally. I saw Bryce as believing deeply that the CEO has the prerogatives that Andrew was exercising. But certainly Bryce was not the only member in this Cult of the CEO. There were two major reasons for the substitution of Ed for Bryce. First, the tightness of the funding environment meant to me that I couldn't have someone on the board who needed funding from the same sources as mine. Bryce had that need, and had convinced both Bud and Fred to invest in another of his ventures. Second, I saw that Ed's skills would be crucial in a sale of the company, which I now viewed as the ultimate way to get rid of Andrew's influence. Any acquirer would have to value the intellectual capital of the company. In my mind, that didn't include Andrew but did include me and the academic connections I brought.

In the board meeting, Andrew asserted he had to devote all his time to product development and that I should find the lead funding sources for the next $2-million round. Bud didn't want to set a valuation. Neither did Fred. He predicted the NASDAQ would fall to 1500, and was unsure if he would participate at any valuation. Ed was new to the board, and didn't see a way to help me out of this bind. Steve and Penny participated by phone, and were as bewildered as I was by this role reversal. With none of the natural sources stepping forward, it was left to me to try to solve this problem.

Ed connected me with the top of Polaris Ventures; Steve connected me with the right person at Koch Ventures. Paul had already hosted discussions with the CEO of one of the major companies in the WPP advertising empire. During the academic quarter break I went off to Park City to ski with my high-school buddies. They dug into this problem with the enthusiasm and optimism only great friends can bring. Mike Grove, a serial entrepreneur who had been up and down in the Denver oil boom-and-bust cycles, worked on an elevator pitch. Rick Mallory, the managing partner of the San Francisco office of the real-estate law firm Allen Madkins Leck Gamble and Mallory, connected me with Val Vaden of Vector Capital and Giffen Ott of Evercore. Mike Jennings wanted to put personal money into the C

Round, but the clients he managed with AXA Advisors didn't invest in such risky ventures. Skip Pennington already had invested in the B Round through his brother Bud, and was planning further investment. I worked on developing contacts at Primedia Ventures and Mayfield.

On April 4 the NASDAQ closed below 1620, and the funds I'd connected with were scrambling to keep their current portfolio companies afloat.

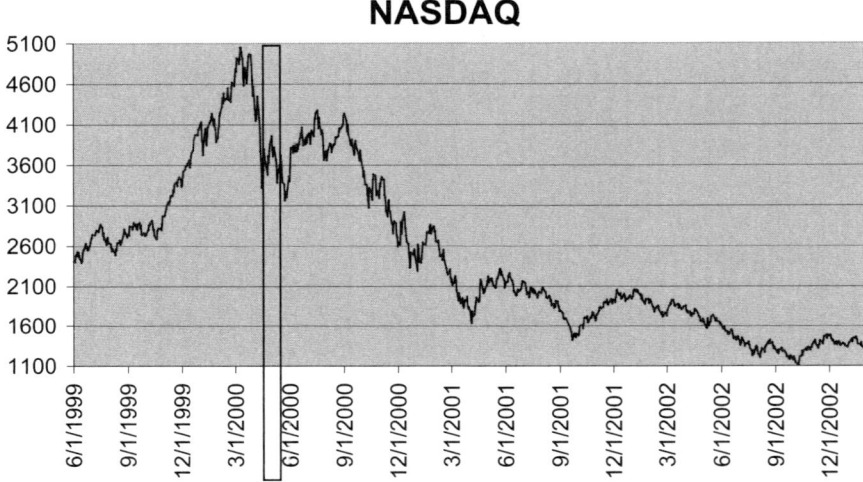

I invited Joe Parks to speak on venture financing to my strategic-marketing-planning class on April 25. After the session he and I sat at Café Roma in the Anderson Quad, and discussed the prospects for the next round. He could broach the topic of C-Round participation with the senior partners only if they had a *full ratchet*. Ah, the dreaded *full ratchet*, re-pricing all of their Series-B shares as if they had been bought at the Series-C per-share price. Even hearing the term sent chills up my spine. "Twist the *ratchet* a notch tighter," the evil Sheriff of Nottingham said as he stretched one of Robin Hood's merry men on the rack to torture him into revealing Robin's Sherwood Forest hideout.[57]

---

[57]Most people remember Basil Rathbone as the ruthless Sir Guy of Gisbourne in the 1938 Errol-Flynn classic "The Adventures of Robin Hood." But Rathbone played Robin Hood in the four record, 78 rpm, Columbia Masterworks (MM583) recording that I listened to as a child. An even more evil voice played the Sheriff of Nottingham.

Despite its prior agreement to a partial ratchet, and its previously expressed desire to participate in future rounds, a down round put Intelligent Technology Ventures in a different mode. The motivation for Intelligent Technology Ventures was more than just getting a better price. With a re-pricing of all its shares, the company's books wouldn't show a loss on this investment. What could make such a Draconian measure worthwhile to me? How about getting rid of Andrew? We talked about possible scenarios that could achieve this, but any final deal would have to have approval from the top people at Intelligent Technology Ventures.

Around this time I met with an old friend and former UCLA doctoral student, Marshall Goldsmith. Twenty-five years earlier I had been on Marshall's doctoral committee and he and I had conducted action-oriented research on citizen involvement in the city-planning process. Marshall now is an enormously successful "executive coach." Recently profiled in the *New Yorker*,[58] his specialty is fixing bad bosses. We met for breakfast the morning after a talk he gave that I, unfortunately, couldn't attend. After describing my current situation, Marshall volunteered his services to try to fix Andrew. I had given up on trying to reform Andrew. I was relatively sure that Andrew wasn't fixable, and certain that Marshall would be viewed as my partisan, despite his impressive credentials. I passed on this very generous offer.

By phone, Marshall introduced me to Jim Belasco, the well-known management consultant and author.[59] Jim had helped Martin Sorell in the dark days of WPP history. Given our discussions with WPP, he offered to speak with Sorell on our behalf. When WPP backed out due to our competition with one of its portfolio companies, I didn't feel it was proper to try this connection to the top. Jim's last advice to me was to "be brave." I'm still trying.

---

[58]MacFarquhar, Larissa (2002), "Whom Do You Call When an Executive is Unbearable?" *New Yorker*, April 22 & 29, 114-136.
[59]Belasco, James A (1990), *Teaching Elephants to Dance: The Manager's Guide to Empowering Change*, New York: Crown. Belasco, James A. and Ralph C. Stayer (1993), *Flight of the Buffalo: Soaring to Excellence, Learning to Let Employees Lead*, New York: Warner Books, Inc. Belasco, James A. and Jerre Stead (1999), *Soaring with the Phoenix: Renewing the vision, Reviving the Spirit, and Recreating the success of Your Company*, New York: Warner Books, Inc.

We began preparing for a May 2 meeting with the principals at Intelligent Technology Ventures. Joe Parks sent spreadsheets that showed what were the implications of a *full ratchet* for the company's part of the Series B and what that would mean if it led the C Round. Already, this was a nuanced change from our discussions that had Intelligent Technology Ventures taking all of the C Round. If it wasn't an alternative to Andrew's participation in this round, the chances of getting rid of Andrew fell with a thud.

After a month of testing on a few campaigns, we finally went live on 11 campaigns at iPlayer.com on April 19. Every day, more campaigns switched over from the learning phase to the optimized execution phase, and our performance improved. We wanted to get the most up-to-date information into the Intelligent Technology Ventures presentation. For the campaigns in optimized execution, we already had a 2× lift. When blended across all the campaigns in learning and execution, the lift was 1.6×, but improving daily.

Intelligent Technology Ventures had relocated from Beverly Hills to the Fisher Pennington building in Century City. Andrew, Ed, Jason, Van, and I represented SDC. David Nalbanian, Joel Wilde, Joe Parks, and Michael Bolton (a new principal partner) represented Intelligent Technology Ventures. (It took me a few minutes to recognize Michael as a student from my marketing-research class perhaps five years earlier.) The presentation went well, as did the Q & A. From Michael's few questions, I sensed he was trying to show me I'd trained him well. I took the incisive questions as a compliment. I attributed Joel Wilde's silence partly to his prior familiarity with us and partly as a sign that any axe he wielded would be used in private.

The crucial exchange was between David Nalbanian and Andrew. Nalbanian asked if Andrew was participating in this round. Andrew said yes. Then, Nalbanian said in that case it was only a matter of valuation. He said that valuation would be somewhere between $5 million and $15 million given the current market condition, and asked Andrew what number that would be. I inserted that the *full ratchet* made the real valuation much less than any stated valuation. We all had reviewed the spreadsheets by that point. I hoped this would provide Andrew the means to argue for a higher nominal valuation. Andrew looked across at Nalbanian and said, "$5 million."

At rare occasions such as this, some part of my mind would float off and consider larger issues, while I would try to focus the rest of my attention on the here and now. I thought: What was the responsibility of a CEO? Wasn't it to maximize shareholder value? Many situations are fraught with the potential for conflict of interest. Rarely do you get such a bald opportunity to witness a CEO putting his interests ahead of those of the company.

The personal side of this was even worse. Instead of splitting $1 million between Giovanni and me for 333,333 founders shares, Andrew had reneged on that initial commitment, run the company into a financial hole, and now was scheming to pay half that amount for around 20 times as many preferred shares.

The conflict was too conspicuous for the deal to advance on the basis Andrew indicated in the meeting. Further, the implied structure treated Intelligent Tech's money more favorably than Bud's money. This was awkward, at best. At the next board meeting, Ed was given the task of being the independent source coming up with an acceptable valuation and deal.

Ed went back to Nalbanian and indicated that, because of Andrew's obvious conflict, a $5 million valuation was not acceptable. After give and take, Ed came back with $7.5 million as a pre-money valuation offer. Since this still implied a different treatment for Intelligent Tech versus Bud and the other investors, problems persisted. Further, Intelligent Tech indicated that it only wanted to put $400,000 into this round. That would mean all its Series-B shares would be re-priced for only $400K in new money. Unacceptable. Intelligent Tech couldn't lead the round with so little money. Ed translated the $7.5 million into what valuation that would imply for an equal treatment for all new money, and came to $3.8 million. This shrinking from nominal to real valuation is what I meant when I interjected my comment between Nalbanian's question and Andrew's reply. Andrew's initial offer in that May 2 meeting implied a valuation of a little more than $2.5 million.

Bud agreed to lead this round at Ed's valuation. Andrew would participate, and Ed even convinced Fred to come into the round. Still, Andrew added three zingers to the terms. The first was a bridge loan. Bud and Andrew would loan the company $150,000 for 30 days, at a 10% annual interest rate, secured by all the assets of the company. This put a gun to my head and, in effect, told me I must sign or give away the company. If the C Round wasn't approved by the shareholders or didn't close for any reason, the cash would rapidly run out, and Bud and Andrew would get all of the assets of the company for their $150,000. The second zinger was that two of the board seats currently controlled by the common shareholders were given to the Series-C investors. This gave the investors five of the seven board seats. So I could give up nominal control of the company or give the company away for $150,000.

The third zinger was a complex performance clause. The round would close as soon as at least $1.5 million of the maximum $2 million for this round was received, but only $1 million would be released on closing. The rest of the funds would become available only if three benchmark conditions were met:
1. For a consecutive, 30-day period, the optimization produced at least 2× lift, including the time spent in learning, compared to a control group composed of these same campaigns. This had to be done for all eligible campaigns – a condition that excluded only a few bulk campaigns that we already knew had to be served too broadly to receive proper optimization.
2. For that same 30-day period, the campaigns used to calculate the lift in (1) had to represent at least 50% of all paid traffic on iPlayer.com.
3. No implementations or tests on other Web sites showed that the positive results achieved on iPlayer.com were not representative of the majority of top 20 online advertising sales Web sites.

These conditions had to be satisfied by August 15 or the rest of the money would be returned to the investors. Presumably, if the conditions weren't satisfied by then, the company would go into bankruptcy, with all the assets going first to the Series-C investors. With all these terms, the round closed on May 30, 2001. The strange added terms scared off a few of the investors. So we closed $175,000 under the proposed $2-million maximum.

On July 8 we had our first chance to meet the percentage trafficking requirement in (2), having finally built the ad volume we optimized up to 52.9% of all eligible traffic. We had total lift (including learning) of 2.06× over the control group. Further, our test campaign for CBS Sportsline.com (using over 8.6 million ads on iPlayer.com) produced

3.27× lift. No failures. I sent all the details to Bud. He asked Ed to check them, and then asked Riordan & McKinzie to release the rest of the C Round to us.

5.4 Last Chance for Strategy

Bud had promised that after the Round C paperwork was completed, Andrew, Bud, and I would get on the same strategic page. I tried to prepare for this my thinking through the strategic issues the way I teach it. I never expected to share the complexities or the strategic map I articulated, but at least I should be clear with myself. The map in Figure 5.1 is, in essence, the way I saw the world.

Figure 5.1. The Mental Map of Factors Affecting SDC's Success

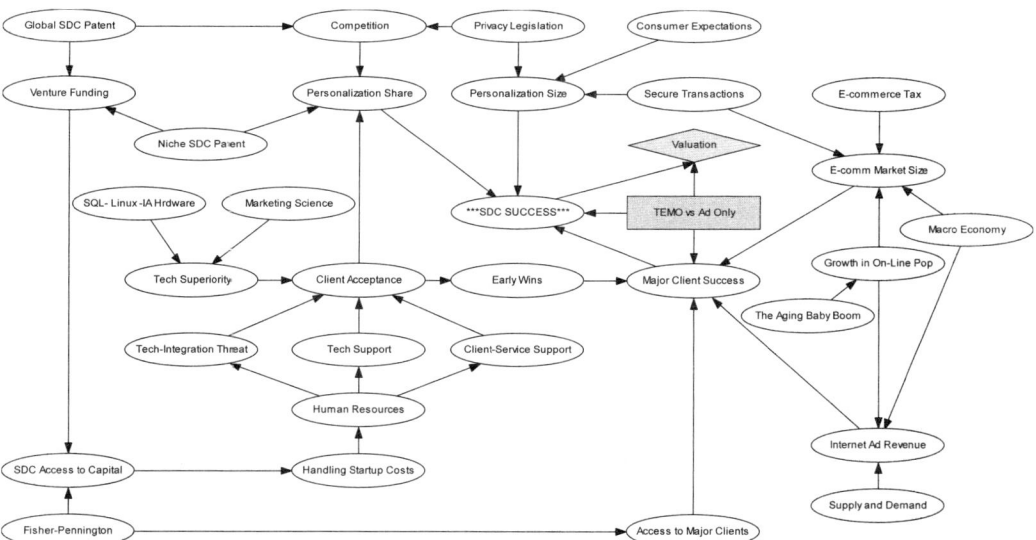

Consider the arrows that point directly to the node "SDC Success." The success of SDC seemed most directly affected by both the size of the personalization market and SDC's share of that market. A big share of a big personalization market leads to a high likelihood of a big win. A small share of a big market or a big share of a small market would more likely lead to a steady state for the company, whereas a small share of a small market would most likely lead to death. Success with major clients would have a direct impact on the success of SDC. These oval nodes are *chance nodes* characterized by answers to general questions such as, "What are the chances that success with major clients leads to overall SDC success?" But the meaning of success depends on the strategic decision between the broader technology enabled marketing optimization (TEMO) focus and the narrower emphasis on Internet ads. Before the crash of the Internet advertising market, the SDC board generally believed that a big win in the advertising arena had more up side than a big win in the broader e-commerce market. I reflected this in the Valuation (the diamond-shaped *utility node*). The rectangular *decision node* (TEMO vs. Ad Only) points at Major Client Success, SDC Success, and Valuation, reflecting the direct impact of this strategic decision on these nodes.

The multi-channel-marketing capability, built into the original design of ZipSegments, gave SDC opportunities for success even if the e-commerce sector suffered decline. SDC's share of the personalization market would be affected by the strength of the competition and the likelihood of SDC obtaining niche patent protection. Global patent protection for SDC would weaken the competition, as would an increased emphasis on privacy legislation. Privacy legislation, along with progress on secure transactions and consumers' expectations for personalization, would affect the size of the personalization market. On the TEMO side, major client success would most likely be affected by access to those major clients, early wins with other referenceable accounts, and the overall size of the e-commerce market. The overall size of the e-commerce market is affected by progress on transaction security, the likelihood of holding off the sales tax on e-commerce transactions, the growth in the online population, and the state of the macro economy. Large-scale demographic shifts, such as the aging of the baby boom, would affect the growth in the online population. Early wins would be the direct result of client acceptance, which also would affect SDC's share of the personalization market. Client acceptance would be directly affected by SDC's technical superiority, tech support, and client-service support, as well as by the perceived threat from technical integration. In different ways, SDC's intellectual capital would drive these factors. Our access to human resources in computer science and management would drive our ability to handle the technical integration threat, staff tech support and client services. It also would drive SDC's technical superiority through mastery of the standard SQL databases, Linux OS and Intel hardware (as long as those standards were stable), and SDC's foundation in marketing science. But the A-list personnel needed for this mastery depended on our ability to handle the start-up costs. Fisher Pennington had a major influence over our access to key clients and capital. Obviously, our desirability to venture funds would affect our access to capital. The venture-fund outlook would be affected by the likelihood of either global or niche patents.

On the Internet ad-market side, the influences discussed above are still at play. The personalization approach works here, since Internet ads are better received if they are anticipated, personal, and relevant.[60] The Internet ad market, however, faced the issue of the supply-demand balance, and the impact of that balance (or lack thereof) on ad revenue. The ability to add advertising is practically unlimited. We know that in the very long term, advertising spending is a mean-reverting process that represents about 2.15% of GDP.[61] That long-term value tolerates wide swings. We also postulate that there is a long-term pressure to shift ad spending to being proportional to the amount of leisure time consumers spend on the Internet. These factors affect the ability of clients to monetize the increase in advertising effectiveness achieved through SDC technology, thus influencing the likelihood of major client success in this space.

The VentureDevelopmentProject.com Website has the complete strategic planning exercise based on SDC. Suffice to say here that all the scenarios I ran indicated the robust strategy was to pursue the broader focus on TEMO rather than narrow focus on Internet advertising.

Just before the mid-July 2001 Board meeting I gathered my thoughts. As Chairman I figured the only discretion I had left was over the agenda of Board meetings. Every time I previously suggested modifications to the Board agenda Andrew broke out in a cold sweat, came up with some alternative, dominated the airtime, and basically got his way. Bud would have limited time, would cut Andrew off, rattle through what was on the top of his mind, and leave. One of

---

[60]Godin, Seth, and Don Peppers (1999), *Permission Marketing: Turning Strangers Into Friends, and Friends into Customers*, New York: Simon & Schuster.
[61]I've heard numerous people speak of mean reversion and this long-term average as well known, but the only explicit reference I've found is: Kornelis, Marcel (2002), "Modeling Advertising Markets Using Time-Series Data" doctoral dissertation, Rikjsuniversiteit, Groningen, The Netherlands, p.115.

Bud's best exit lines was when he left the Board meeting, which we usually held in the Fisher Pennington conference room, saying the President of the United States was waiting for him upstairs. He was referring to Martin Sheen.

The meeting Bud promised to get on the same strategic page never happened. So at the upcoming board meeting I would insist on a discussion of strategy, despite Andrew declaring, as he huffed out of my office, that inserting this into the agenda meant "total war" between us. I prepared the following statement that I presented in a Chairman's Report to the board members:

> **Strategic Decision Corp Situation Analysis: The Chairman's Viewpoint**
>
> **SDC Value Proposition**
> Strategic Decision Corp employs technology-enabled marketing to translate our clients' strategic customer data into monetizeable assets.
>
> **SDC Strategic Assets**
> **PersonalClerk Ad Advisor** – This ad-recommendation engine finds the customers most interested in each ad offering as evidenced by the doubling of click rates across a broad diversity of ad campaigns. The base hardware configuration works at 1,300 recommendations a second with an architecture that is completely scalable to client needs.
> **PersonalClerk Product Advisor** delivers real-time product recommendations that are tailored to each customer's preferences. PersonalClerk Product Advisor starts targeting at the first visit and immediately learns about preferences to further refine the product recommendations it delivers.
> **ZipSegments** – This segmentation scheme is the only multi-channel direct marketing tool actionable for targeting and personalization on the Web, in email, for mobile messaging, and for snail mail. It ties U.S. customers to the tremendous resources of the Census.
> **Intellectual capital in the science and practice of modern marketing.** SDC's connection to the best minds in technology-enabled marketing truly give us the ability to fulfill our motto "From Analysis to Action" with practical and scalable methods.

As the emphasis shifts from pure click-rate maximization to branding and conversion, we understand how our technologies support modern marketing best practices.

AT Kearney and others have shown that the missing link in supply-chain integration is proper quantification and automation of the demand-forecast component. We understand how our combination of traditional statistical modeling and datamining places SDC in an advantageous position with respect to this cross-industry trend.

As new data streams become available to track emerging commerce arenas our collective expertise can harness these data to the advantage of our clients and us. Just as the analysis behind ZipSegments launched our PersonalClerk suite, the ability to translate analyses into scalable products is a strategic asset for SDC.

Strategic Weaknesses:
Sales: We have not translated our iPlayer.com effort into appropriate revenue.
Sales: We have delayed contact with Yahoo!, AOL, and Microsoft.
Sales: We have not moved fast enough with CBS Sportsline.com, Pogo.com, and other Doubleclick accounts.
Sales: We have been too hesitant to pursue revenue opportunities with (Studio X) Home Video Division, Bear Stearns Japan, and others.

Environmental Analysis
The business environment has undergone wrenching change in the last 15 months. The early days revealed an environment in which capital was readily available. Both the retail and advertising applications of our e-commerce suite (PersonalClerk) seemed to have well-funded and adventurous clients. We have witnessed the deterioration of the pure dotcom e-retail environment. The Internet advertising market experienced a precipitous decline, only now showing signs of the beginning of recovery. With rare exception the survivors are firms pursuing hybrid strategies:

Bricks-and-mortar firms that are cautiously adding Internet advertising to their overall advertising portfolio – no longer in a mad rush to establish an Internet presence, no longer in fear of being left out.

> *Bricks-and-clicks firms* that use their digital presence primarily for product information, aftermarket support, and as part of a multi-channel customer acquisition strategy, while using their physical plant for distribution logistics.
> *Pure dotcoms* that build multiple revenue streams from advertising, user fees, e-commerce transactions, etc. These firms have learned the traditional lessons of *owning* their customers, and moving them down the path of awareness, interest, evaluation, purchase, and purchase-event feedback.
>
> **Five key issues influence all in this business ecosystem**. Their impact on SDC is indicated below:
> **Privacy**: While our basic approach is privacy friendly, extensions of our technology are somewhat vulnerable. Focus on conversion from clicks to registrations or purchases require the passing of identified information across Web sites. This must be handled by acceptable bilateral agreements between the publisher site and the advertiser site.
> **Personalization**: This is developing as the norm for successful customer experience on the Web. We gain in such an environment, but so do any potential competitors.
> **Price competition**: While commoditization is still a threat to all of e-commerce, emphasis on personalization and branding both mute price competition and favor SDC.
> **Peer-to-peer communication**: This trend has little direct bearing on SDC.
> **Patents**: We are in a strong patent position. The U.S. Patent Office has granted our "Application to Make Special." This means our patent application will be reviewed shortly.
>
> **Best Practices**
> Our approach to technology-enabled marketing works. We have sustained more than 2X lift for five weeks. The goal now must be to find the market(s) that value our fundamental offering. Historically, firms in our position *fail* if they do not accept the limitations of the new technology. Find a fertile niche and then pursue incremental improvements.

I tried to gain a little more attention and air time by relating that I was soon leaving for Washington to receive an award from the American Marketing Association and the Marketing Science Institute for my work in strategic marketing planning for radically new products. Unlike most academic awards, I pointed out that this one was given for outstanding contribution to the *practice* of marketing. This prelude helped me get five minutes of uninterrupted time to present my position.

Fred Hart had often been the most mercurial member of the board, so I focused my remarks toward him. He seemed very impressed with the statement, bought into the value proposition, and agreed with the statement of best practices. So did Bud, Steve, Penny, and Ed. Andrew remained silent. After the meeting I set about confirming our understanding in an email that iterated the basic points of agreement.

> From: Lee Cooper
> Sent: Friday, July 20, 2001 2:03 PM
> To: EPinter@xxxx.yyy; SMayer@xxxx.yyy; FredHart@xxxx.yyy; penny_baron@xxxx.yyy; AHarper@xxxx.yyy; BudP@xxxx.yyy
> Cc: David VanArsdale; Jason Kapp; Lee Cooper; PaulB@xxxx.yyy
> Subject: Thanks
>
> Thank you all for your support of SDC in the board meeting.
> I'm very pleased that we achieved clarity on the value proposition we offer: "Strategic Decision Corp employs technology-enabled marketing to translate our clients' strategic customer data into monetizeable assets."
>
> I'm also pleased that we understand our core strategic assets:
>
> PersonalClerk Ad Advisor,
>
> PersonalClerk Product Advisor,
>
> ZipSegments,
>
> Intellectual capital in the science and practice of modern marketing.
>
> These are the assets that underscore our ability to live up to our motto: "From Analysis to Action."

> I understand the tough tactical position we must execute, but feel the company benefited from the agreement on the strategic underpinnings of our efforts.
>
> Cheers,
>
> Lee

Andrew reacted as if the discussion never occurred:

> "Andrew Harper" 07/20/01 08:13PM
>
> Hi Lee,
>
> I do not understand what you are saying here. If this has a meaning regarding how we run the business, I do not understand it nor did the board endorse anything. Please clarify this so I understand why you sent this email and what it means to you.
>
> Thanks,
>
> Andrew

I tried to put the role of board decision in a context Andrew could not deny.

> Hi Andrew,
>   When the board discusses matters of strategy or policy it is our responsibility to air any differences, come to an agreement, and abide by the sense of the outcome. Thus, if Fred indicates that we must come to a resolution of the option re-pricing issues that has no negative tax implications for the employees, we are bound to follow that. Similarly, if you disagree with the SDC value proposition specifically discussed or the list of the strategic assets of SDC, you should discuss these differences, come to a resolution, and proceed.
>   There are some specific meanings for what we do in running the business, as well as general guidelines that help us make day-to-day decisions. In the Yahoo! presentation we should replace the "SDC's Objective" on page 20 of the board packet with the SDC value proposition. In concluding our contracts with members of the SDC advisory panel we should remember that

> maintaining good relations with the marketing science community is a strategic asset of the firm. And, as Fred indicated, we should look at non-Internet applications of our methods as aligned with the strategy of our firm.
>
> Cheers,
> Lee

Andrew's reaction was to ignore this board agreement and completely remove funding for the Office of Research and my administrative support. That will teach me to mess with King Andrew the Specious. Unless the board was willing to fire him, which it wasn't, he had total power.

5.5 D is for Doom

The summer and fall were characterized by technical success and market failure. Using the existing technology, SDC optimized over banner ads for iPlayer.com and lifted performance 2.7× for more than 2 billion ad impressions. We ran a few CPA (Cost per Action) tests with clients on iPlayer.com and obtained lifts from 3.6 to 3.9×. In these tests, involving 60 million ad impressions, the clients let us put a one-pixel icon on their purchase page so that we could carefully track the people who actually purchased a product after seeing and clicking a banner ad on iPlayer.com.

Bud had arranged for high-level meetings with MSN and Yahoo! I warned Bud that this was a dual sale – first to the executive level, and then to the heads of the technical staff. Without expertise on the marketing-science side of our efforts, we would never pass due diligence. This was particularly true with Yahoo! where the responsible technical head held a PhD in management science and operations research. I was ready for his questions, but no one else in the company was, in my opinion. I repeated this warning, leading up to several client trips. Bud always agreed, then talked to Andrew, and I was excluded from the meetings. King Andrew could not be upstaged.

While a lot of excuses came back -- complaining of Yahoo!'s not-invented-here syndrome, for example -- the overarching lesson is that good technical people can tell when someone is faking it. They asked for white papers; Andrew sent them marketing copy. The executive level was always impressed by the profit implications, but half a sale isn't good enough.

Andrew expected all of this to change once his technical approach to ad serving was put in place in early December. His Semi-Real-Time (SRT) method was an outgrowth of the incoherent ramblings he had produced in the early fall of 2000. My treating him like "a failed doctoral student" had apparently created the *idée fixe* of proving me wrong. Since spring, he had the tech team trying to develop his SRT method into something workable. It was such a radical departure from the traditional forecast-and-budget approach that backed the baseline code that I had repeatedly warned against attempting this second radically new technology. I had initially assumed that the benchmarks he was preparing to propose for the C Round would be a stalking horse for the SRT method. Thus, in May I wrote concerning the yet-to-be-revealed benchmarks:

> From: Lee Cooper
> Sent: Friday, May 11, 2001 7:59 AM
> To: AHarper@xxxx.yyy;
> Subject: Benchmark
>
> Andrew,
> What do you propose?
> The best practices in this arena are very well established.* This is the critical juncture when a firm offering a radically new technology must find the market that values the core technological offering. The eagerness and enthusiasm of the iPlayer.com sales staff indicates that the CPC arena is our entry point. CPC becomes the foot-in-the-door for our CPM business. Sustaining innovations (incremental improvements) are always possible unless the resources required to achieve those improvements starve the marketing-finding efforts. But spending the next two months chasing another technology breakthrough is a certain path to failure.
>
> This is go-to-market time!
>
> *Bower and Christensen (1995), "Disruptive Technologies: Catching the Wave," *Harvard Business Review*, January-February, 44-53. (HBR Reprint No. 95103).

> Clayton Christensen (1997), *The Innovators Dilemma: When New Technologies Cause Great Firms to Fail*, Boston: Harvard Business School Press.

Andrew denied that there were any best practices in this arena. Whatever plot was behind the benchmarks clause was foiled when the existing technology performed beyond threshold. This didn't dissuade Andrew from his SRT mission.

As reported earlier, almost daily for three months Andrew had dismissed successful simulations with the Share-of-Choice (SOC) method. He used the technology team during that time to create something potentially workable out of his Semi-Real-Time (SRT) method. By June, the SRT method was ready for testing and we agreed to a bake off among the existing Baseline method, the SRT, and the SOC method.

Constructing a valid test should not have been hard. The two recognized forms of validity are construct validity (i.e., does your test measure what it is supposed to measure?) and predictive validity (i.e., does performance on your test predict performance on some outside criterion?). In our case, the criterion selected was the performance of the live system. This was encapsulated in five days of live performance of 12 ads.

The major departure in the simulator from the actual performance on the live system was that Andrew arranged for these campaigns to receive traffic volumes that were grossly higher than they had in the live system. Such a tweak substantially benefited the SRT method, since it relied much more on heuristic, within-day adjustments than on what has proven over the prior 100 days to be a simple but effective daily-econometric forecast. The econometric forecast took a day or two to adjust to these new volumes. In a 30-day normal cycle, this is minor; in the five-day simulation, it was more of a handicap. Despite this tilting of the playing field, I continued with the test.

Every laboratory scientist knows that when it comes to measurement, the first task is to calibrate your instruments. In our case this meant, "Does the code that generated the results in real life generate the same results in the simulator?" For this simulator the answer was "No." The baseline code produces a weighted lift (i.e., the click-through rate of the live campaigns compared to the same campaigns in the simulator) of around 1.7, while a calibrated simulator should produce 1.0. Thus, I felt it was important to calibrate the simulator, to align it with the live system. If it did align, we could expect the performance of any other methods tested in the simulator to be predictive of their performance on the live system. If the calibration revealed the simulator to be overly optimistic (weighted lifts over 1.0), we could re-weight the other methods tested to get a more accurate forecast of their performance on the live system. To obtain this accurate calibration, one must run the simulator with exactly the code, configuration file, and supporting tables as used to generate the live results. Andrew refused. It was as if moving to what outside experts would call a *fair test* moved the results too far from his control.

Even in the face of these obstacles, I reported to the management meeting on July 9, 2001, the results summarized in Figure 5.2 below:

Figure 5.2. "Bake-Off Results (Static Simulator).

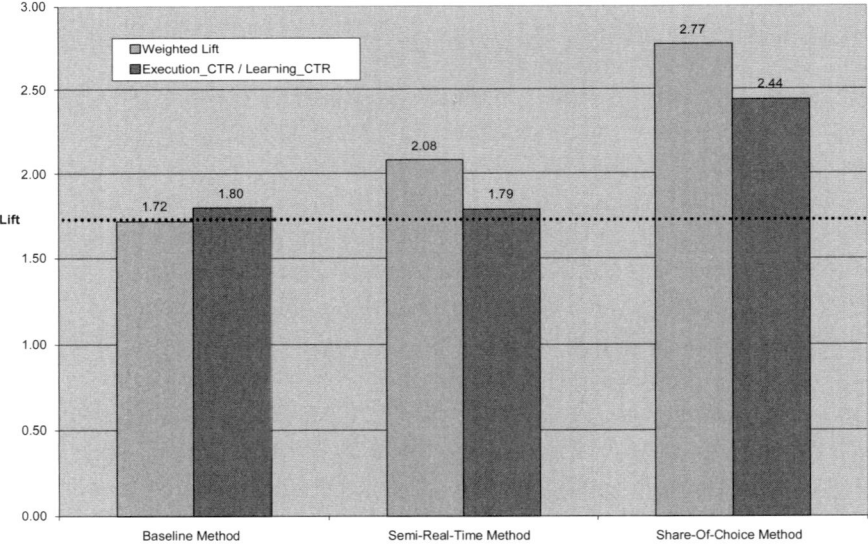

Over 4.5 million ad impressions, there was no need to calculate if these differences were statistically significant; they were. After a few days of diddling with the numbers, Andrew's response was that we had to change the answers in the simulator and start over.

At this point, Xuegao realized that Andrew was never going to implement anything other than the SRT method. He started returning the phone calls from companies that wanted his services. Andrew acted here as if the only way he'd take a test would be by first reading the answers in the back of the book. He'd run a simulation with the new answers and adjust the internal parameters of the SRT method until he got the result he wanted.

As Andrew kept running simulations to fine-tune his answers to match the back of the book, I became more and more convinced he would never do a final test. I posted an SOC result that was 126% over the baseline method with a lift of 4.8×. Andrew was still trying the read the answers at the back of the book, so I began running fake simulations to try to lure him into competing. My mental image was that of a sailboat jockeying for position before a race start – feinting and jibing to confuse the opponent into misreading the intended starting strategy. His best simulation of the SRT method yielded 129% of the corresponding baseline method, but a 4.6× lift. So Andrew averaged all my fake results together with the real one, which, he said, gave the SRT method a 10%-20% advantage, and declared himself winner of the bake-off. I pointed out that the SRT method did worse than the control group in around 48% of the traffic. That would be terribly irksome to the advertisers that were de-optimized. The SOC method de-optimized on around 1% of the traffic. This didn't matter to Andrew. He won. Xuegao left. I was stuck.

Andrew's penchant for defining reality reminded me of that ominous passage from *Darkness at Noon*, Koestler's chilling novel based on the Moscow Trials of 1936-38.[62]

> Extracted from the diary of N. S. Rubashov, on the firth day of his imprisonment …
> "A short time ago, our leading agriculturalist, B., was shot with thirty of his collaborators because he maintained the opinion that nitrate artificial manure was superior to potash. No. 1 is all for potash; therefore B., and the thirty had to be liquidated as saboteurs … if No. 1 was in the right, history will absolve him. If he was wrong …"

---

[62]Koestler, Arthur (1941), *Darkness at Noon*, New York: The Macmillan Company.

Andrew used the scalability requirements introduced by MSN to push his agenda. So in early December, despite Bud's repeated statements that we would go to Yahoo!, MSN, and AOL with the existing (baseline method) technology, Andrew went live with the SRT code. But, as the simulations indicated, the SRT method de-optimized some campaigns -- doing worse than the control group. When the overall performance was below the control group, Jason quietly started taking these bad campaigns out of optimization – wiping out their historical performance, and artificially inflating the apparent overall performance. Still, the system never again quite achieved the 2.7× lift the baseline code delivered in the first 2 billion ads.

By mid-November, I knew my contract would not be renewed beyond its December 31 expiration. I wrote good-bye emails to all of the employees and shared warm follow-up exchanges with many of them. I intended to continue my symbolic role on the board of directors until I saw the terms of the D-Round financing.

Andrew announced severe personnel cuts on December 1. Only nine key people would remain, plus Andrew. Still, the cash would run out before the end of January.

Ed was again designated by the board to come up with the best deal he could. Given that Andrew had run the company up against the wall, and delayed seeking financing until the last minute, we could expect no funding outside of existing investors. The terms were largely Andrew's to dictate. Four of the seven board seats would be designated to the D Round. The common shares would get one board seat. Since employee options were rendered largely worthless by this round of funding, Andrew proposed cash bonuses to key employees if profitability benchmarks were met. Of the potential $585,000 in bonuses for the whole company, $250,000 was reserved for Andrew himself. The shares would be prices around $.08 each, but would get a three-fold liquidation preference before other investors saw a penny. The D Round would get drag-along rights that could force a sale of the company even against the wishes of earlier investors. And, of course, a bridge loan was put in place, so that I had another gun to my head. Agree to these terms, or the whole company will be turned over for only $50,000.

Steve and I decided to resign from the board. The NASDAQ was below 1900 by the beginning of December. Most technology companies were simply not receiving the next round of funding.

Still, I had to decide whether I would sign the agreement. I saw the company as doomed on its current path. Then I thought about those people who remained. This round of funding should last the company until June of 2002. That would give Giovanni enough time to finish his PhD dissertation. His wife, Catarina, should also finish her dissertation by then, as would Jason's wife Sara. Fabrizio's girlfriend at that time was finishing her MBA at NYU, and planned to move out to L.A. by June. Guiseppe was just married. Chuck Yu had married not all that long before, and was thinking of starting a family. I decided I would not sacrifice their employment for whatever ego gratification I would get from refusing to sign the D-Round papers. I signed.

My board term was set to expire at the end of January 2002. I decided to wait on my formal resignation until the appropriate time.

Clearing out my SDC office was a nostalgic exercise. The December 1 lay-offs had left the sixth floor largely empty. Operations would soon consolidate on the second floor to save rent. The idea of writing this book began to gel, and I fantasized writing about the people who once occupied these now-empty cubicles and offices, of the meetings and arguments that flashed around the conference table, of the hopes and schemes that had dominated my mind for the prior 30 months. I wondered if anyone would listen to the lessons when they were punctuated by my failures. Success is much easier to sell.

When I returned on January 2 to pick up my last paycheck, Van said Andrew had instructed him to send me to see Andrew in my old office. Andrew, sitting in my old chair behind my former desk, just looked at me and said, "You can get in line like any other vendor." I told him he was in breach of my contract. I said, "You're an egocentric fool," turned and left.

Technically, 30 days had to pass before I could file a suit for breach of contract. Classes were starting again. I was set to teach strategic marketing planning in the digital economy, and starting to write this book. I waited until January 11 to fax in my Stockholders Response and Consent to the Series-D stock offering. In spite of the breach of contract, refusing to sign this would be against my interests. But I resolved to sign no other documents, including my board resignation, until the breach was resolved.

The closing was delayed from the intended January 15 date because the attorneys wanted to include the new board structure. In my mind this was not a necessity, rather a convenience to the attorneys. I wouldn't resign until the breach was remedied.

On Tuesday, January 22, I spent all morning chairing a meeting of the ISRC, the UCLA committee that handles conflicts of interest in sponsored research. I returned to my UCLA office shortly after lunch to find a message from Ed. He had been tasked with making the closing happen. When I returned his call, he told me they were very anxious to close on Wednesday. He said that he would get Andrew to promise in writing that he would make the final payment within 24 hours of the closing. I said if he would fax that letter to my study at home, I would fax back my resignation.

Now I had to decide my immediate course of action. I needed new tennis socks, and had planned to find some before going back to my study. Should I rush back to deal with the closing or take the long route by Santa Monica Place and shop for tennis socks first? What do you *think* I did?

5.6   *Kiretsu Versus Portfolio*

The Website VentureDevelopmentProject.com contains a strategic planning exercise for Strategic Decision Corp. The central issue in the exercise concerns the decision to emphasize Internet advertising as the corporate focus or engage more broadly in technology enabled marketing optimization (TEMO). After laying out the map of all the major forces impacting SDC's market potential, the likely payoffs can be determined under a broad range of assumptions. The bottom line on that exercise is that in the best of all possible worlds, pursuing Internet advertising has a greater upside for SDC. Nothing ever proceeds exactly as fantasized. If everything goes as in the baseline plan, both options have approximately equal value. In the face of a wide range of uncertainties, the TEMO approach has a much better chance of paying off. The optimal decision from the planning efforts described above would be to follow the more robust path of technology-enabled marketing optimization (TEMO). Why isn't this the path that the company followed?

It is too easy to assert that the egocentric Andrew realized that the TEMO approach would require my active involvement in technology development and bridging to the academic community. I believe this was true and unacceptable to Andrew, but more was at play.

SDC was part of a Kiretsu that was led by Fisher Pennington. If it were simply a part of a portfolio of companies or investments, portfolio theory makes the same recommendation as those of traditional test theory.[63] If you wish a test to cover an intellectual domain, you gather a group of items that are validly related to that domain and have a modest negative correlation with each other. That way, you cover the breadth of a domain with partial indicators of knowledge of the underlying domain. The same logic applies to combining judgments from experts. You don't want five experts who always agree. That is no better than one expert. You want a small number of people of equivalent expertise (like items of equivalent validity) who tend to disagree somewhat over time (a modest negative correlation). If you wish to spread risk over a series of investments, you select those that have positive correlation with the investment criterion (profit measures) while having modest negative correlation with other investments in the portfolio. This maximizes profit over the portfolio for a given level of risk tolerance.

In a portfolio you want investments seeking the same goal from different routes. So if one is down, not all of them are likely to be down. A modest negative correlation between asset returns ensures this. In a Kiretsu, the individual companies may appear independent, but they are subservient to the goals of the leader. In SDC's case, the interests of Fisher Pennington had much more to do with advertising than e-commerce or technology-enabled marketing. Fisher Pennington's other clients were content producers. The overriding question was, How would creative content be supported in the Internet (i.e., digital) world. SDC's value to Fisher Pennington was related to how we helped Fisher Pennington answer that question. Andrew's interests in advertising meshed with Fisher Pennington's

---

[63]Ghiselli, E.E., Campbell, J.P., & Zedeck, S. (1981), *Measurement Theory for the Behavioral Sciences*, New York: Freeman.

interests. My interests in technology-enabled marketing and the future of marketing science were overmatched.

So SDC did not fail to plan. Despite my reluctance to share my mental map in its entirety with the board, the major conclusions were repeatedly articulated. What SDC failed to do was take the more likely path to success – the more robust path. Instead, it followed the riskier path that served both Andrew's interest in advertising and control and Fisher Pennington's interest in advertising. While Bud was enthusiastic about personalization and its potential in e-commerce, success on the advertising side of SDC was what helped Bud's other ventures the most. In the convergence of the Internet, entertainment, and computing, the major issue is how content/creative efforts get valued. The Hollywood/content side drove Bud's thriving practice. If an advertising model could succeed in supporting Web content, Bud's other business interests would benefit greatly. What I didn't understand when I allowed Andrew in as CEO was that he was interested only in the advertising side. Thus, Bud's interests and Andrew's focus were aligned from day one.

5.7   *The Tale of DVX*

Why did Bud continue to invest in SDC? Perhaps because the Kiretsu interests were still being served, Bud (and Andrew) continued to invest in the company, and SDC survived when so many other companies shut down. Yes, Bud didn't want to be seen as disserting the investors he brought in. Yes, he was the pivotal power on the board, and so he might experience giving up on the company as a personal defeat. When he welcomed me to the firm's clientele, he did brag that every company in the Internet practice had been a success. Time had changed that. I knew of just three major business failures in the Kiretsu since that bold statement, one of which had a $100 million valuation crumble to zero. Certainly, by the D Round, his historic relationship with me had nothing to do with it. I think, however, that more than personal loyalties or bragging rights were involved. I may never know, just as I'm sure Bud will never tell me what private deals he had with Andrew. But my best guess is that his tenacity in this deal came from lessons he learned from his experience with DVX. What follows is what I've heard of that tale.

Part of Paul Brendl's responsibility at Fisher Pennington involved understanding the technology frontier facing the entertainment industry. When, in the early 1990s, the industry was forecasting that 80% of U.S. households would have broadband access sufficient to carry video-on-demand by 1999, Paul knew that forecast had to be wrong, which of course it was. He set out to find out whether anyone was working on an interim technology. Nobody was. He and Bud talked about the possibility of using a strongly encrypted version of DVD as a delivery vehicle for movie rental. If the encryption scheme was rock solid, a dial-up connection would allow rental, re-rental, or purchase. Production costs on the DVDs were cheap enough that the DVD need not be returned after rental. Further inquiry revealed that nobody was trying to develop this. The next obvious question: was no one doing this because it couldn't be done? Bud and his partner, Ken Fisher, put a little money into engineering studies. It could be done. Since their firm had the studio connections that would help ensure content for the new rental medium, they decided to develop the technology they named DVX.

About $5 million split between the two of them produced a working prototype and the organization needed to take it further, with Paul as president of the nascent company. Bud received encouraging signals from the major studios, except Warner Bros., who seemed determined to control their film library themselves. Bud and Paul began shopping DVX around to the major electronics retailers. WalMart was very interested, but candidly admitted that it was not an early adopter. DVX would need another champion, but WalMart wanted to follow. The champion turned out to be Frederick Cutter, the CEO of a leading national retailer (over 600 outlets) of brand-name consumer electronics, personal computers, movies, music, and games. He agreed to repay Bud and Ken their $5 million, and fund the next $30 million round for half the equity in the startup.

Known as a savvy businessman and vicious competitor, Cutter was also known for his desire for control. All of the checkout POS systems in this retailer were of its own design. Many other retailers used off-the-shelf systems, while this one developed custom applications. When the time of the expected next round of investment approached, the CEO seemed hesitant to commit to the $75 million price tag. Bud took him at his word and began discussions with other sources, primarily the software mogul Gil Bates. Bates was interested, perhaps seeing strategic advantage in the Hollywood connection, and decided to fly to L.A. to meet with Bud and the DVX top staff. Somehow, Bud felt that he had to tell Cutter about the upcoming meeting. Cutter immediately arranged to fly out. Bates left the meeting thinking that a deal was at hand. But Cutter started working on Bud, agreeing to provide the entire $75 million. This was Bud's strategic inflection point.

When I've retold this story in my strategic marketing planning class, I've stopped at this point and asked what the MBA students would do and why. Almost everyone suggests taking the Bates deal and backs the decision with one of 10 good reasons. The trump card, however, is control. The Bates deal seemed perfect. He would put in the $75 million for a one-third interest, balancing Cutter's control. The Fisher Pennington one-third stake would be able to decide any conflict between these two gorillas. What went on in the private meeting between Bud and Cutter, I'll never know. But somehow Bud decided that Cutter would provide the funds, making his retail chain the majority shareholder in DVX.

The real problems began when the DVX team tried to line up other retail partners to distribute the hardware. The retailers understood the value of forward pricing the hardware to get enough installed base for the software. They were certain that Cutter would undercut the software prices, despite assurances to the contrary. When Paul would explain how it wasn't in Cutter's interest to do this, other retailers would remind him of the tale of the scorpion. To them, Frederick Cutter was a scorpion, and whether or not it was in his interest he simply couldn't resist an opportunity to screw a competitor.

The upshot was that Cutter's retail chain was almost the sole retail site pushing DVX. Even with this limitation, the Christmas 1998 ad campaign for DVX was a success. When I met with Bud in January 1999, he indicated the sales figures were ahead of forecast.

Within three months Cutter unceremoniously shut it down. A total of 1,500 people were fired. Bud woke up one morning with paper wealth $60 million dollars less than the day before.

Why? The demand was there. The service was operational. The content poured out on schedule, and rental demand was strong. Two factors: The forward pricing of hardware put an immediate drag on the retailer's gross margins. The more hardware it sold, the worse for the short-term stock price. This was happening at a time when a number of senior managers were about to vest their retirement portfolios, which short-term bonuses would affect greatly. They whispered in Cutter's ear that DVX was hurting corporate valuations.

The solution was to spread the cost of forward pricing while augmenting the installed base by encouraging other retail electronic stores to sell the hardware. Cutter, the scorpion, could not help his competitors, even if it would help him.

I have since heard that the other studios that had held out were just about to sign on to the DVX standard. The result would have forced the reluctant retailers to get on board. Bud's potential $60 million could have turned into a much larger amount of *real* money.

The lesson for Bud, I think, was never to give up control. Continuing to invest in SDC was the result of that expensive and hard-learned lesson.

5.8   E is for Epilogue

I had seen the end coming the previous summer, and embarked on a program to undo the damage that my experience had wrought: physical, emotional, and academic.

On the physical side, my exam for key-personnel insurance in June of 2000 revealed triglycerides more than twice the level that starts the "high" range, total cholesterol and LDL (bad) cholesterol far beyond warning levels, and high blood pressure. I didn't need an exam to tell me I was almost 20 pounds over my target weight. I don't remember giving permission for further tests, but was also informed that I was HIV negative, with no traces of cocaine, marijuana, or nicotine. Insurance companies are very nosy.

I procrastinated until August of 2001 before seeing my own physician, trying my own hand at better diet and exercise. By then my triglycerides had fallen 100 points, but remained far above the level defined as high. My cholesterol was down a little, but was still far too high; my blood pressure remained elevated; I'd lost maybe five pounds; and I now showed up as glucose intolerant – perhaps headed for type 2 diabetes. My physician from the sports-medicine division at UCLA expressed serious concern. He started me on appropriate medications, set up visits with a dietician, and order me to get 45 minutes of brisk exercise every day. During the previous 18 months I had been lucky if I squeezed in one tennis match a week.

Ann and I used to walk almost daily. Starting again would be good. I found modern diets, with their emphasis on fruits, nuts, and good oils, to be actually more interesting than the low-fat, high-carb diet I supposedly followed. When I returned for a follow-up in early November, I proudly reported that I'd been walking 35 minutes each morning, and had only missed two days in the prior six weeks. My doctor's response was, "I said 45 minutes." When I protested that I already got up at 6 a.m., he replied, "Get up 10 minutes earlier." Apparently that was the kind of sympathetic response I needed. At my last checkup my triglycerides and cholesterol were both in acceptable ranges, glucose was below the warning level, blood pressure was normal, and I had lost half of the extra pounds.

I mention this partly because many people feel that academics are counter-dependent on any authority figure. I certainly was critical of the SDC CEO, the dean, and numerous others in positions of authority. But when authority was backed by knowledge and sound judgment, I had no abstract difficulty following it. The practical difficulties of being disciplined about diet and exercise notwithstanding, my willingness to follow my doctor's orders helped me see that my conflicts with Andrew were not merely a function of counter-dependence.

On the emotional side, by the end of 2001 I was down, but not crushed. In many ways my strategic inflection point came in the fall of 2000, when I decided that I couldn't sever my ties to UCLA. Leaving UCLA had never been my plan – but the crisis of that fall had raised the possibility. The four walls of my faculty office enclose a sanctuary – one that survivors of the academic promotion process value as a haven of accomplishment. If a faculty member is content there, so be it. I had wanted to venture out – to be in both worlds. I wanted the academy *and* I wanted the real world. I felt the university and the real world both gained when faculty venture out beyond the ivory tower. My sense of failure came from not being able to have both. But half a loaf is better than none. That may sound glib, but I was returning to a life that had been a source of fulfillment for many years.

So I was wounded, and I believed that writing about my experience would help heal the wounds. But I couldn't write out of a sense of loss and bitterness. Not until I realized how special the summer of 2000 was in my life and the life of the newly born company did I have an angle that could communicate what was lost. I thought about the college students: Alex, Nick, Brandon, Jonathan, Daniel, Jeff, Matt, and Eric, and how important this first job was in shaping what they expect from the careers ahead of them. I thought about the technology team, and the special feeling we had that summer as we brought the technology live in the iPlayer.com test. That's where I could start. At the end of January, I began to write in earnest.

On the academic side, my hopes of shaping the next generation of marketing-science models were gone. I correctly assessed that I would need infrastructure to support the huge data-handling requirements, and lots of funds to support colleagues in their efforts. My days of building huge SAS programs were over.

What did I want to do? I only wanted to write. Why did I want to write? Partly to tell this tale, partly to help other academic entrepreneurs avoid some of the traps I fell into, and partly to tell the business world that a great opportunity is being missed. The wizards inside the academy and the wizards in the business world don't know how to talk to each other. I hoped to write something that could help find a common language.

I figured out a way to accelerate my sabbatical, normally not due for another year, and began writing seriously. And that is what I've been doing ever since.

Almost.

In early May, 2002, I began hearing rumors about an E Round. My initial reaction was that it was time to shut SDC down. I expected some further screw job from Andrew. If there was nothing in this round for the common shareholders, I wanted to ensure that I had a straight up or down vote on the company's future. All I initially asked from Ed, who had stayed on the board to represent the common shareholders, was that no bridge loan be structured this time.

Ed said that Andrew was talking about a deal with Banner Market Place (BMP) – a division of a large media conglomerate that dealt with cost-per-click ads -- that would make SDC cash-flow positive by July. Despite my long-held view that the cost-per-click side of the business was where our core technology was valued,[64] I was skeptical of anything Andrew said at this point. Ed indicated that he would talk with Van about the deal, to assess how real it was. I said I'd talk with Jason to see what was up from a more technical point of view.

Jason told me that three deals were in the works: BMP, which was close to signing; eGalaxy, which was similar technologically; and a division of DoubleClick, for which we were doing an alternative to its ClickBooster service. All the deals involved CPC, which was good. None of these clients had the customer data that drove SDC's traditional core technology. On the one hand this was bad, since the basic model works best when it employs ZipSegments to learn about customer preferences. On the other hand it was good -- if SDC could make money without customer data, Andrew had less of a chance of screwing it up. The deals amounted to doing smart ad trafficking in a CPC environment. The technology team had learned a lot about trafficking in the last 18 months. Given that you have an opportunity to offer an ad, what is the most profitable ad to offer under the existing wear-out conditions with the current diversity of inventory? SDC gets paid only if someone clicks. BMP deals with the large-format pop-up and pop-under ads that annoy so many people…but are noticed. Baseline click rates for these ads are much, much larger than those for the typical banner ad.

Ed and I compared notes. It seemed real. I wondered, If it was only a matter of cash to cover the period between billing and actually receiving cash, why couldn't SDC get a loan to cover accounts receivable? Ed indicated it might be possible to factor the receivables, but it would be costly and would require establishing banking relations beyond our current arrangements. That added time, and time was getting short.

The deal that was being offered was essentially the same as the D Round. The D and E shareholders would jointly designate the four board seats. Liquidation preference would go to the E shareholders with a three-times payout before anyone else got money. The E Round would get drag-along rights that could force a sale even against the wishes of earlier investors.

---

[64] As I indicated in the May 11, 2001 email presented earlier.

This financing round was coming about just when I was writing about some of my most horrid experiences with Andrew. It was as if my self-therapy was interrupted by an unwelcome intrusion of good news. I took my frustration out somewhat on Ed, whining my way through one phone call on why, under these circumstances, we could not get a better valuation. Penny got analogous treatment when she phoned with positive news out of the May board meeting. She had gotten them to skip the bridge loan, and refresh the pool of options for employees. I was sure Andrew was constructing some vile scheme.

When a neutral or mildly positive stimulus (the content in Ed's and Penny's phone calls) evoked a disproportionately negative response (my constructed conspiratorial theories), I knew I had to work on gaining perspective and emotional clarity. I had already removed the ongoing irritant, having resigned more than four months before. By writing, I had taken the opportunity to bring structure to my understanding of what I had been through. I could partition my sense of loss into the part that was my own doing, and the part that was done to me. When separated in this manner, I felt I could handle the regret over my part. Regarding the rest, I just had to face the brutal facts and put them behind me.

This is really not *that kind* of a self-help book, but the steps in this process were pretty clear:
- Disproportionate responses are a signal that something is off.
- If the situation can't be remedied, remove yourself.
- Use whatever help you need to gain a better cognitive understanding of what went on. For me, that was writing, but other tools suit other people.
- Partition the pain into what you brought on yourself and what is attributable to others.
- Find some way to handle your own part. For me, the partitioning made it easier to simply accept responsibility for my part.
- Face the brutal facts of what's left.
- Move on.

Good friends stick with you in such times. Ed, Penny, and Steve were there for me. Ed had constructed the best deal he could wangle out of Andrew. Penny had gotten me the up-or-down vote I wanted. All three had helped talk me through the dark side of this experience.

Now that I had my up-or-down vote, what would I do? First, I apologized to Ed and Penny. Second, I decided to approve the deal. And third, I considered whether I should exercise my pre-emptive rights in this round.

Say I owned 10% of the shares of the company: My pre-emptive rights allowed me to buy 10% of this round, so that my overall percentage ownership would not be diluted. At around $.08 a share with a three-fold liquidation preference, it looked like a good deal. That wasn't enough. We still believed a strategic sale would be the exit strategy, with DoubleClick as the obvious target. If the company sold for $11 million or less (including bankruptcy), none of the common shareholders, including me, would get anything unless they participated in this round.

To protect myself against a low-price sale, I started investigating. I ended up informally interviewing everyone in the company, except Andrew. I held a long discussion with Giovanni, who had decided to invest in this round for much the same reason I was thinking about. His relationship with Andrew was OK. In many ways, since the tension was relaxed, he gained from the exchange of ideas. Daily Ad Caps didn't apply in the CPC side of the business. Once the horrendous constraints of the Daily Ad Caps were removed, Andrew could no longer micro-manage random events, and Fabrizio was able to code the trafficking rules very quickly. Guiseppe and Chuck were productive and well deployed. Ravi was central to both the technology and the management. They felt these three clients could be brought on line with no additional technical staff. In fact, they believed that adding to staff would decrease their productivity, given the steep learning curve. Jason was general management with an emphasis on Client Services. Brad was Jason's only remaining Client Services representative. He felt quite able to handle both the BMP and the eGalaxy accounts. The work with DoubleClick involved posting results to an FTP site. No Client Services involvement was contemplated for the early round of integration.

While only Giovanni and I, as founding shareholders, had pre-emptive rights, the other employees were being let into this round, if they wished, as a partial fix for having their options so far under water. A number intended to take the offer.

The conservative management summary, sent with the Notice to Investors, listed two signed contracts and one agreement that shortly turned into a signed contract. The near-term cash-flow projections on these ranged from $135,000 to $245,000, while the monthly burn rate was closer to $110,000.

This business would have competitors until SDC could move its clients up the services ladder and convince them to start using customer data. Then SDC would have more competitive insulation. But SDC was there, with signed contracts, and that gave SDC a temporary advantage.

SDC identified six other potential clients in this area whose combined volume equaled that of the three top players already under contract. I could easily fantasize $3-6 million in free cash flow. In a growth sector, which this is, that could translate into a sale of the company for $30-60 million. That's a far cry from the $1 billion fantasy that Andrew conned us with, but it still looked good to me.

I decided to exercise my pre-emptive rights.

The bad news is that the $3-6 million in free cash flow hasn't yet materialized. Through the spring of 2003 the company flirted with profitability, but anticipated some cash flow problems in the early summer as new clients were being brought up to speed. The cash flow problems were handled through very short-term loans at minimal interest rates from Pennington and Harper. August 2003 proved to be the first month of substantial profitability, and September billings pointed to the company's first quarter of profitability.

The nice part of value-based pricing is that SDC can see how much money its technology is making for its clients. SDC brings in demonstrated incremental profits that fluctuate between $300K and $500K per month for its largest client. BMP's exclusive rights in its competitive arena expired in May 2003. From that point, the prospects for SDC start looking a lot rosier. BMP's competitors could see that BMP was able to profitably serve ads that they could not. SDC management translated that advantage into new clients that began to come on line in the early summer. BMP is no longer SDC's largest client. The backlog of clients interested in deploying SDC's technology has grown sufficiently that the company finally can secure some upfront, development payments for implementation. SDC has also been able to prove the robustness and scalability of its hardware design. Its new largest client served more than 1 billion ads in August 2003 on a system of SDC's design that integrated 30 Web servers.

Perhaps one day I will actually be able to profit from what I helped create.

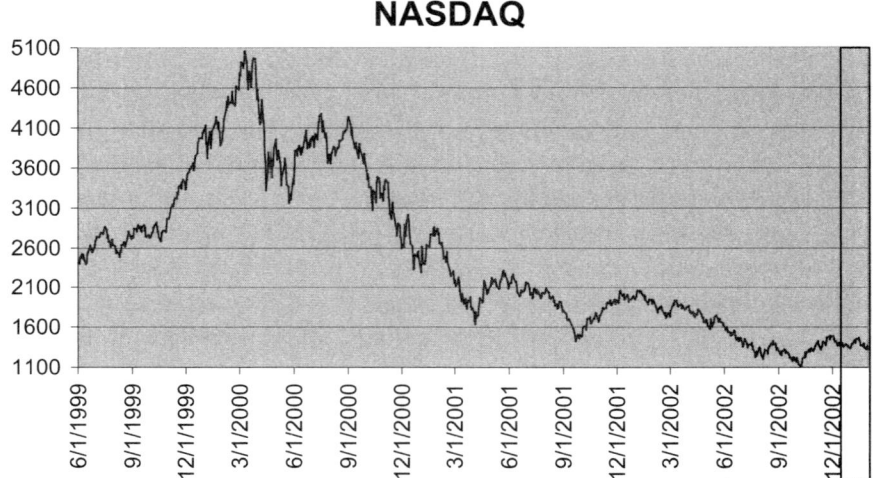

PART II. RECONSTRUCTED LOGIC

6. A Linear Path

This chapter presents the steps involved in moving a radical innovation toward commercialization as if a linear path could be followed.

6.1 Introduction to the Linear Path

As I indicated in the opening section of the book, starting a new venture doesn't follow a linear path; I create this convenient fiction to help teach some of the steps. A true-to-life path will be somewhere between a straight line and a series of circles. But remember that the only sustainable strategy in the face of constant change is one that recognizes where you are and knows, consequently, the proper next step.

I begin with the innovative idea and the understanding that a radical breakthrough can impact many markets. Understanding the kernel of that idea from a business perspective can lead to aligning the innovation with the best first market. I follow this with a discussion of the value of the entrepreneurial vision. Understanding the business kernel of the innovation can lead to a clearly communicable vision of the venture that becomes a powerful tool for the entrepreneur. With these two building blocks in place, I begin a discussion of writing a business plan, as distinct from planning a business. I conclude the current chapter with a discussion of the due diligence you must expect whenever a financial decision is deliberated.

Writing a business plan is done for a temporal set of reasons, for a specific audience, addressing a particular set of questions. Planning a new business is inter-temporal, dynamic, and meant to reveal and simultaneously address the complex set of issues facing that new venture. Chapter 7 deals with this topic. Chapter 8 concludes with discussion of the more general lessons I've learned from this venture.

6.2   Kernel Analysis: Aligning Innovations with Markets

Radical innovations, particularly those that are the products of university labs, do not necessarily come with area of application clearly delineated. Radical innovations can find application in many different markets. The basic proposition behind successful commercialization of radical innovation is that you must find the market that fundamentally values the kernel of the innovation.[65] Then you trick out the kernel just enough to give that market a compelling reason to buy.[66] Sustaining innovations follow, along with fame and fortune.

In trying to think of how to illustrate this, I've often fallen back on Steve Mayer's invention of the programmable video game, the Atari 2600. Mayer and Larry Emmons, his partner in Atari predecessor Arvin Electronics, ran the advanced engineering division from the founding of Atari. Until Mayer's innovation, the coin-op business dominated, and video games had essentially a single-purpose functioning – one box, one game, with each game being engineered from scratch. When the quarters stopped flowing, the box was replaced by the next dedicated machine. Atari arcade games were rapidly copied, shortening the product life cycle even more severely that normal. Nolan Bushnell's strategy was not to sue for copyright or patent infringement. Rather, he responded to the "jackals" by constantly generating new games and ideas. By 1974, Atari was producing a new game every other month.[67] The demands on Mayer and his group for custom designs were enormous. He needed to come up with some dramatic simplification if the engineering of new games was to proceed on schedule. Mayer somehow understood that the kernel was a hardware platform that combined a display manager with graphics and instruction processors. The game software was content that tricked out this kernel to appeal to popular tastes.

---

[65]C. M. Christensen, "The Innovators Dilemma: When New Technologies Cause Great Firms to Fail," Boston: *Harvard Business School Press*, 1997.
[66]G. Moore, "Inside the Tornado: Marketing Strategies from Silicon Valley's Cutting Edge," New York: *Harper Business*, 1995.
[67]S.L. Kent, "The Ultimate History of Video Games," New York: *Prima Publishing*, 2001, p.61.

I wondered where such great insights come from, so I called Mayer and asked – expecting some ethereal equivalent of "Eureka." He said, without a moment's hesitation, "The HP-35."

Introduced in 1972, the HP-35 sold for $395, which converts to more than $1,740 in CPI-adjusted 2002 dollars. It used a multi-chip CPU. The Control and Timing (C&T) chip performed all the major non-arithmetic functions, including generating instruction addresses sent to the ROM. The Arithmetic and Register (A&R) chip contained seven 56 bit (14 BCD digits) registers – three of which were the X, Y, and Z registers I remember struggling to program to perform simple statistical operations.[68] The kernel of the innovation that Mayer recognized was the separation of the basic functioning of the system in the CPU chips from the ROM chips that contained the instructions for the particular functionality the HP-35 offered. The HP-35 used three ROM chips that contained 256 instructions of 10 bits each. To go from the HP-35 functionality to the HP-45 financial calculator required little more than a different set of ROM chips – at least, that was the rumor that Mayer used for his insight. He never looked inside. When we recently looked out of curiosity, we saw many more changes than just the ROMs. Sometimes the rumor distills the innovation better than the physical reality.

Perhaps nothing new exists under the sun – all "new" things being merely reinventions. Even if that's true, genius exists in recognizing that prior genius can be migrated to the current context. When Atari needed to go rapidly from one game to the next, Mayer saw as the answer a design that separated the general-purpose computer from the content/software component.

Recognizing the kernel of innovation is not an easy task. Two substantial barriers stand in the way. First, we need to learn how to translate from the world of innovation to the world of business. And second, the business side must resist the temptation of extremes: to see the innovation as either changing the world, or as a useless dalliance. Business leaders and venture capitalists must learn how to ask the questions that lead to finding the right market – the one that fundamentally values the kernel of the innovation. They need to understand where markets and innovations intersect to shape the

---

[68]See http://www.hpmuseum.org/techclas.htm .

minimal augmentation of the kernel that provides a compelling reason to buy.

From a business perspective, the kernel might not be immediately obvious to the innovator/technologist. In the case of the Atari 2600, other major design innovations were candidates for the kernel. What I called the general-purpose computer actually contained two components. The true general-purpose component was the MOS Technology 6502 CPU that later drove the Apple I.[69] Mayer recognized that the 6502 was not capable of handling the whole display. But he figured that if he designed a display co-processor to handle the horizontal aspect that ran at 15 Hertz, the 6502 had enough power to drive the vertical aspect of the display that ran at 60 Hertz.[70] While this required a little more clever programming, it had the great side benefit of reducing the display memory requirement from a function of display area to a function of display height plus display width. This, combined with his fundamental patent for moving an object across a video screen in response to a joystick, allowed 8-bit objects to move anywhere across a 32-bit background. Even greater creativity was needed to design the operating system, the game play, and the graphics into the 128 bytes of RAM the original unit contained. (That's bytes, not kilobytes.) Though Mayer later won an Emmy for technical achievement for his design of the DF/X Composium, a digital post-production suite for Mayer's post-Atari startup Digital F/X, he describes programming the 128 bytes in the precursor to the Atari 2600 as the most difficult engineering challenge he has faced.

The original units were stand-alone games, such as the Pong game, but the internal design included the innovation that led immediately to the Atari 2600 video-game system. Later, 2-kilobyte and (still later) 4-kilobyte cartridges for the 2600 contained both the game and the operating system. There was no native operating system, so each game cartridge defined the complete software environment. The rudimentary OS could be tweaked as required to play the individual game without worrying about legacy compatibility.

---

[69]Steve Wozniak worked on the development of the HP-35. He sold one of the prototypes of the HP-35 to provide some of the seed money needed to start Apple.
[70]The special processing chip for the horizontal aspect was called *Stella*. The team had the practice of naming their chips after girlfriends of team members. When this engineer's turn came, he was without a girlfriend, and so this famous chip was named for his bicycle.

One characteristic of the kernel is that it can provide the ability for the technology to go beyond the vision of the technologist. Mayer had designed the system to handle the kinds of games he then had in mind. In addition to *Pong*, the team designed the first maze game, *Gotcha*, and the first racing game, *Trak 10*. These were all games played on two-dimensional grids. Mayer had no idea the market would lead him into three-dimensional spaces with whole new game scenarios. But with the kernel he had created, Atari could respond to the sometimes-almost-overwhelming demand.

The other side of the problem comes from the business world's failure to ask the right questions. Both in boom times, when seemingly any proposal that includes the buzzwords of the moment can get funded, and in lean years, when even likely winners have difficulty attracting backing, the appropriate questions are not being asked. Some opportunities are squandered, while others are missed completely.

The three questions to ask are:
1. What is the business kernel of the innovation?
2. What is the market that fundamentally values this kernel -- who cares?
3. What is the minimum augmentation of the kernel required to generate a compelling reason to buy?

In Atari's case, the tacit appreciation of the kernel (i.e., separating the hardware support for all games from the software/content of a particular game environment) led the company to billions of dollars of annual sales during its heyday. The Atari game players created an essentially new category of leisure-time activity. While there is no denying its roots in the coin-op-game business, the home-video-game category could not have naturally evolved from the nightspots and game arcades that first featured Pong. The kernel of the innovation (i.e., the separation of content and hardware that enabled game cartridges) was more valuable in a consumer product – one that required very different channels of distribution from the B2B coin-op business. The prospect of profits from cartridge sales allowed forward pricing of the hardware console; that is, early consoles with much higher unit costs could be sold at a loss to build the installed base. As the demand for hardware consoles grew, the cost saving from production volumes would bring the console price closer to profitability. It took the initiative of Gene Lipkin, Atari vice president of sales and marketing, to find a helpful phone operator in the Sears Tower, someone who just kept trying different people until finding the receptive ear of Tom Quinn, the sporting goods buyer. His decision to devote the back page of the 1975 Christmas catalog to Atari involved an entrepreneurial leap of faith for which Sears is rarely credited. That single act put Atari into the consumer business, first with the minimum augmentation needed to generate a compelling reason to buy – the stand-alone Pong game – and then with a market eager to buy the Atari game console and cartridges that followed.

In the case of HP's revolutionary 1.3" HD Kittyhawk drive, the kernel was clear at the beginning (i.e., to build a small, dumb, cheap hard drive). The internal HP funding the Kittyhawk team secured, and the high-level support that generated those funds, attracted the top engineers -- all wanting to be part of the next hot project. And that original mandate got lost in the exuberance of gathering a hot design team. The Kittyhawk team identified several markets that cared (e.g., storage for Nintendo game players, and the PDA market exemplified by Dayton Electronics). The question the team failed to answer was the third, "What is the minimum augmentation of the kernel required to generate a compelling reason to buy?" At second contact, Nintendo approached the Kittyhawk team with a complete system, with a slot prepared for plugging in the drive, if HP could meet the $50 price point. But the engineers were more set on solving the PDA-markets occult engineering-design problems, such as tolerating a three-foot fall without data loss, than mundane design-for-manufacturing problems.[71] The proprietary six-axis piezo-electric accelerometer fails to meet anyone's criterion for the minimum augmentation of the kernel. The result was that the team had one shot at finding the market, and insufficient resources to try a second time, when the right initial market had become clearer to them.

Mayer's second startup, Digital F/X, began with a simple vision of distributed video: design a video-editing suite affordable to hobbyists wanting to enhance home movies. Mayer had a clear idea of the minimum feature set needed to accomplish this goal. A year of searching for venture backing ended when Kleiner Perkins agreed to fund the effort. As mentioned in Chapter 3, this was Vinod Koshla's

---

[71] J.L. Bower and C.M. Christensen, "Disruptive Technologies: Catching the Wave," Harvard Business Review, (January-February 1995), 44-53.

first deal after leaving Sun Microsystems. Mayer acknowledged that bringing in Kleiner Perkins meant turning control of the company over to others. His vision was supplanted with a much more ambitious plan to redefine the state of the art. All of the hottest chip designers wanted to be a part. Instead of the simple, inexpensive unit Mayer had initially envisioned, his team pushed the envelope. The first two units were nicknamed Mickey and Minnie, and shipped off to Disney. But so much of the development budget was spent on tricking out the kernel to meet the heavy demands of this premier customer that the resources were not available to do the mass-market machine. As mentioned above, Mayer won the Emmy for technical achievement, but the company did not survive.

The late 1990s are rife with stories of venture capitalists seizing on promising innovations and trying to change the world, funding huge leaps and land grabs, when seeking the minimum augmentation needed to find a market would have been more prudent for the funds' investors as well as the innovators.

6.2.1 Finding the Kernel

The basic premise is that there are many ways that the kernel can be tricked out to provide a market with a compelling reason to buy. That implies that the kernel is what makes an innovation translate into many different applications possibly serving many different markets, and is what makes the innovation an application platform. Thus, the separation of hardware from content in Atari games enables the rapid introduction of new games, meaning that profits from games can partially subsidize the hardware costs, meaning that more hardware platforms can be sold at the lower price point. The virtuous cycle this describes produced great profits in the early Atari times. For the HP pocket calculators, the kernel was also the separation of the hardware platform from the ROM that provided the functionality for different applications (e.g., the HP35, HP45, etc.).

Understanding the kernel is also fundamentally important in maintaining strategic control of how value migrates in a value network.[72] Prior to Atari's innovation, profits (particularly in the coin-op business) were associated with machines — that combination of

---

[72]See A.J. Slywotzky, "Value Migration: How to Think Several Moves Ahead of the Competition." Boston: Harvard Business School Press, 1996.

unique hardware and software that made up a game. After Atari's innovation, profits migrated out of the hardware into the software. Atari's lack of control of the software that ran on its game player, while creating a generation of game programmers, contributed greatly to the firm's demise. A wave of inferior third-party games diluted the franchise. Atari's own bad bets on blockbuster titles (e.g., the ET game) that turned out to be flops sealed its fate. The lesson of value migration was not lost on Nintendo in the next generation of video-game companies. Nintendo controlled tightly what titles were released for its player.

The Venture Development Project in the Price Center for Entrepreneurial Studies at UCLA is attempting to use these lessons to help guide the migration of UCLA innovations toward the marketplace. One effort involves Core Micro Solution Systems (CMSS), a startup established by Professor C.J. Kim in the micro-fluidics arena. Most applications in the MEMS (micro-electrical mechanical systems) arena use clever devices for miniaturizing mechanical systems for special purposes. Thus, in micro-fluidics, miniature pipes and pumps move fluids in a tiny version of freeway traffic, where overpasses and underpasses ensure proper direction of payloads. Like the coin-op machines of the early Atari, a different freeway system is designed for each application. Professor Kim's approach, however, recognizes that at micro- and nano-scale, the forces of adhesion and surface tension are stronger than the forces of weight. Thus, we have the scene in the movie *Antz* in which an ant caught inside a drop of water struggles to get out, barely able to burst through the surface. Creatures smaller than an ant do not have the skeletal strength to overcome such surface forces. At this scale, bubbles of air, if properly controlled, function as valves, and electrical forces can act as pumps.

The basic functionality of Prof. Kim's innovation consists of the ability to:
1. Separate a drop from a reservoir.
2. Move a drop along a path.
3. Merge one drop with another.
4. Divide a drop into two parts.
5. Move separate drops independently on a two-dimensional grid.

Electro-wetting on dielectrics (EWOD) is the technology that drives these functions for droplets currently around 10 nanoliters in size. What makes these simple functions a platform for applications is that each of them can be controlled electrically, by programming, rather than by special-purpose micro-manufacturing. The kernel of the innovation is in the program control: using the fundamental properties of surface tension and adhesion at nano-scale to create programmable, discrete droplets that move independently over a dielectric surface.

We are still investigating what is the best first market for the platform technology, but a strong early candidate is the high-throughput-screening market. This market, currently about $1.7 billion per year and experiencing robust growth, needs precisely the five functions listed above to move samples and reagents into the proper grid spots for analysis.[73] The prototype device is remotely programmable using the IR port on a Palm-Pilot-like device. Maintaining the programmability as a separate component means the grid can be inexpensively manufactured and disposable after each use.

With disruptive innovations such as electro-wetting on dielectrics, many potential market applications are possible. This technology has applications possible in areas as varied as optical switching, inkjet printing, fuel cells, sample preparation for DNA chips, and biohazard monitoring field stations. Finding the right first market is key to securing the resources needed to grow. Because of the engineering challenges associated with chip fouling, longevity, and evaporation, we believed that high-throughput screening presented the best first market. Long-term growth, however, is likely to occur by applying the technology behind the innovation in another market.

It is very important to understand the implications of the multiple and varied markets that radical innovations could capture. A deep understanding of the technological possibilities must be maintained at the top decision-making levels of this startup. Venture capital firms like to put experienced management into place, which is fine as long as the experience is not simply bound by market knowledge. New ventures need experienced managers and market knowledge. Determining where in the organizational hierarchy these needs are

---

[73]S. Fox; "Fine-tuning the technology strategies for lead finding", Drug Discovery World, Summer 2002. S. Fox, "High-throughput Screening 2002-New Strategies and Technologies," High Tech Business Decisions, Inc., 2002.

satisfied requires very clear, long-term perspectives on the diffusion of innovations. Top management has to understand the migration path for the innovation. Market knowledge should come into the venture below the CEO level, as business-development managers for example. If a CEO is brought in for his or her market knowledge in one sector, the venture may never get to the other markets where the innovation may have its greatest impact.

Taking into account what faculty innovators want in terms of desired roles and continuing involvement as their innovations move toward market is the key to designing technology-transfer programs that are productive and sustainable. Many faculty interested in commercializing their innovations are quite willing to learn new things – to become something more than what they were. Very few are interested in becoming something else. When choices are posed as "either-or," faculty stay closer to ivory-tower roles. When choices as posed as "and," I find much greater faculty willingness to add skills to their repertoire – learning how to translate the language of the laboratory into the language of the marketplace.

6.2.2   Market Finding

Which market *cares* is not always obvious. What guidance is available for finding the market that values the kernel? Some basic questions can help guide the search.

Can you hear the voice of the market? The 40-year-old "new marketing concept" is that manufacturers should listen to the voice of the customers and provide the products and services those customers need. In many places in this book, I've talked about Clayton Christensen's thesis of how established companies listen to the needs of their best customers, allocate resources to meet those needs, and get run out of business by radically new products from unknown companies. Am I changing my tune now? No. This is more about how those unknown companies find radical solutions that fit the pain articulated in the persistent problems of various marketplaces.

How can university-based innovators follow this advice? First, you have to listen rather than talk. Effort such as the California Nanosystems Institute (CNSI) frequently bring distinguished speakers to present research ideas and findings to the intellectual community. But industry-university meetings usually feature university speakers doing more talking than listening. When we do listen, it is more likely we are listening to each other, rather than voices from the marketplace. I'm certainly guilty of this. So I think CNSI and other such consortia should hold small format sessions where industry people talk about the problems they cannot solve or the inefficiencies they regularly confront. The hope is that someone will recognize how a 5-percent change in a current project can create a major opportunity for addressing pressing problems in a particular marketplace.

Is the innovation a disruption in an established category? If so, the firm needs to look carefully at the *value network* in that category to seek co-evolutionary partners. The *supply chain* describes the series activities that get raw materials and subassemblies into a manufacturing operation smoothly and economically. The *value chain* is the generalization of the *supply chain* that looks at every step from raw materials to the eventual end-user -- right down to disposing of the packaging after use -- with the goal of delivering maximum value to the end user for the least-possible total cost. The *value network* is the generalization of the *value chain*. Rather than a linear chain, an interdependent *value network* of firms and trans-firm standards are needed to supply a whole-product solution. The value network of the early video-game-console business was spelled out above. By the time of the Nintendo generation of video games, this network had expanded to include many other partners, with storage capacity as an important need.

One simple summary of the value network can be obtained from looking at dollar expenditures in a broadly defined market. Our initial market guess for the micro-fluidic technology from CMSS was high-throughput screening (HTS). The drug-discovery market of which high-throughput screening is a component is shown below:

162 | Midlife ~~Crisis~~ Startup

Figure 6.1. The Drug-Discovery Market

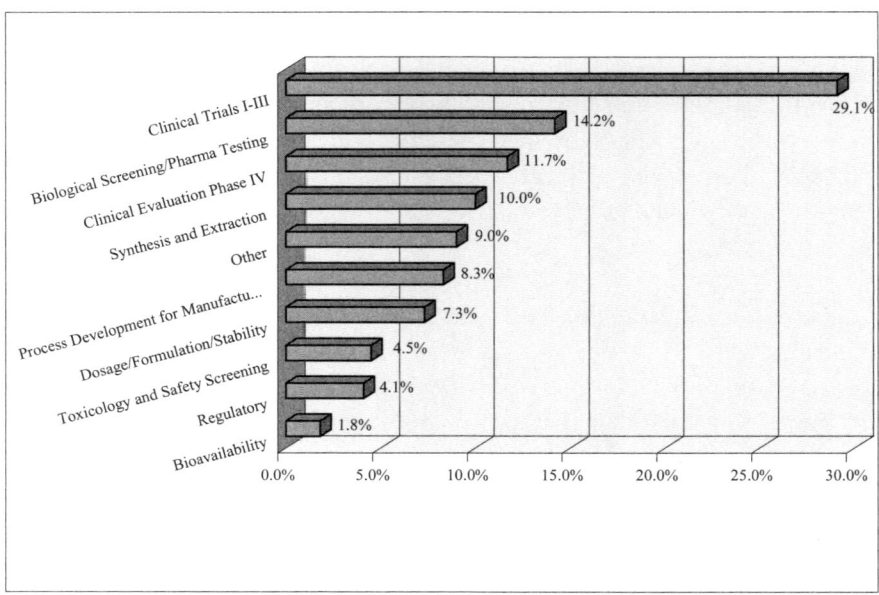

HTS accounts for the major part of the biological screening and pharmacological testing budget for drug discovery. HTS is also becoming involved in toxicology and safety testing and bioavailability. As noted before, the pharmaceutical and biotechnology industries are aggressively pursuing improved drug discovery enabling technologies. Since HTS has proven very successful at discovering more leads for new drug candidates, this portion of the value network is very important. The market drivers pushing HTS include: intense competition among pharmaceutical and biotechnology companies to find new targets, new drug leads, and new drugs; pressure on drug companies to reduce the time to market for new drugs; pressure on drug companies to consolidate activities and reduce costs; increased spending by drug companies on R&D for drug discovery; and increased pressure to screen more targets with higher-throughput and more information content while keeping costs low. The overall market for HTS in 2001 was estimated to be worth $1.7 billion (Fox 2002b). The student team estimated very conservatively that CMSS could impact at least $453 million of the HTS market in 2003. The team expected this figure to grow to $638 million by 2010, assuming all environmental factors remain relatively the same.

In summary, articulating the value network in established categories provides good opportunities to size the market, when the innovation impacts existing categories.

Is the innovation a disruption that creates a new category? The Atari game players created an essentially new category of leisure-time activity. This always makes analysis tougher and more speculative. But successful products fill needs. Specific needs are nested within more general needs. So, while Dean Kamen's Segway™ Human Transporter may establish a new category of personal transportation devices, it clearly falls within the existing, more general need for transportation. Similarly, the Atari game player created a new category, but that category fell into the more general need category of leisure-time activities that for retailing purposes included sporting goods, games, and toys. So when a disruptive innovation creates a new category, the venture needs to look at the next more general level of category to understand its competitors. At that higher level, most players will look like competitors, and finding co-evolutionary partners may be less likely. Channel partners may exist, either in the form of consulting companies and/or service providers that help support the innovation after sale, or in traditional retail or wholesale distributors trying to maximize the use of their assets.

What are the enabling characteristics of the innovation? It may be no accident that the examples that fit most clearly as disruptions in established categories are components, while ones cited as creating new categories are more like finished products. But more is operating here than a simple part-whole distinction. On one hand, it is just a matter of how generally you conceptualize the need being served. A whole product in the storage category is a component in the computer-systems category, and a whole product in the home-video-game category is a component of an overall leisure-time-utilization category. On the other hand, a disruptive innovation creates a new category because it enables products or services that previously didn't exist.

The enabling characteristics in the kernel of the innovation change the value equation in some important way. For Atari, the kernel radically changed the time and costs of creating new products. Soft-ionization techniques are enabling mass spectrometers that cut weight (and size) by a factor of 50, vacuum requirements by a factor of 20, cost by a factor of 10, and run on a 9-volt battery. This combination will take mass spectrometry out of the laboratory, lead to a generation of mobile sensor-analyzers that can monitor our water and air supplies, and help ensure the safety of our mass transportation systems.

What are consumers' expectations? Application of nanotechnology to textile can create widely differing enabling characteristics. On one hand, nanotechnology can produce textiles that are stain resistant and water repellant by their very nature – attributes long desired and valued by consumers. On the other hand, nanotechnology can create shirts that contain computer keyboards, or monitors for heart function – attributes never evoked when consumers are asked what they expect from shirt fabric. Obviously, the steps in marketing a radically new product that enables desired but previously unattainable attributes is much different from marketing toward needs consumers never knew they had. In the overall consumer decision cycle of awareness, interest, evaluation, intention, purchase, and purchase feedback, the stain- and water-resistant fabric can be marketed by emphasizing these attributes in the evaluation stage. Marketing to previously unknown needs requires basic work in the awareness and interest stages. Understanding the enabling characteristics of the innovation is fundamental to making sound business decisions.

6.3   *The Value of the Entrepreneurial Vision*

The entrepreneurial vision creates a mission for the organization. In SDC's case, the vision involved technology-enabled marketing that fulfilled the dream of personalization of the Internet experience and set the foundation for the future of marketing science in Web-mediated enterprise. The early employees *got it*. Whether or not they could grasp the structure behind the models and methods, they understood the context that this vision created for their efforts. To me, this is what Haeckel (1999) means by *organizational context*:

> A well-articulated context provides an unambiguous framework for individual activity, aligning and bounding organizational actions without dictating what those actions should be. It leaves empowered individuals free to choose the best response to unanticipated requests within a unifying

framework of unambiguous purpose, principles, and structure.[74]

Jason *got it* through his understanding of how SDC filled the gap in the IT supply-chain analysis that AT Kearney undertook. He easily spread that vision through SDC's Client Services organization partly because Troy and Carol also *got it* – Troy through his prior work with me in digital convergence and strategic planning in turbulent environments, and Carol through both her MBA curriculum and her prior efforts in helping MBA students deal with computer technology. Kate *got it* because it is a natural extension of modern-marketing principles (e.g., segmenting, targeting, and positioning) to Web contexts that she had practiced in bringing banks online. Ravi *got it* long before taking my class "Marketing Strategy in the Digital Economy," and once introduced to it there, couldn't wait to become a part of the effort. Van was the first person I tried to explain it to, and his ease in *getting it* was key to my sense that this vision could be clearly communicated.

All six of these key early hires hold MBAs from the Anderson School at UCLA. The marketing curriculum they needed to *get it* could have come from any of the 40 schools that think they have top-20 MBA programs. How much the shared culture of the Anderson School facilitated the sharing of vision, I cannot say for sure. I'm more convinced that once they got the vision, the Anderson experience helped them act as empowered individuals in a team effort.

Is entrepreneurial vision a prerequisite for a successful company? Obviously not. One of the main points of the Collins and Porras (1994) study that underlies their management classic *Built to Last* was that visionary companies, rather than visionary leaders, seemed more fundamental to long-term excellence.[75] Note, however, that the time frame they chose essentially eliminated modern startups. True, they inquired about the presence of visionary leaders at the earliest stages of long-enduring companies. But by looking at companies more than 50 years old, they focused on industries that were vertically structured, rather than the more specialized horizontal network structure of modern economic webs.

---

[74]Haeckel, Stephan H. (1999), *Adaptive Enterprise: Creating and Leading Sense-and-Respond Organizations*, Boston: HBS Press, p.17.
[75]Collins, James C. and Jerry I. Porras (1994), *Built to Last: Successful Habits of Visionary Companies*, New York: HarperBusiness.

To me, having a clear entrepreneurial vision before the company gets off the ground can be a great accelerator of progress. This is the elevator pitch, the jump-on-the-stage 30-second speech, and the calmly spoken opening that leads to deeper communications.

6.4   Writing a Business Plan

Writing a business plan is different from planning a business. A *plan* for a business is a dynamic framework that assesses the complex problems, issues, and opportunities facing a venture, and sets strategy for the future. This is the subject of the next chapter. The document titled *Business Plan* is a temporal document that communicates to a specific audience what that audience wants or needs to know about the venture in order to make some decision(s) concerning the venture. To understand how to write such a document you need to ask: What time is it? Who is the audience? What decisions will that audience make based on this business plan?

Asking, "*What time is it?* " highlights two issues: *what time is it* in the history of the venture and *what time is it* relative to macro-economic cycles. This explicitly recognizes that what you are expected to know about your venture changes rapidly over the early life, and this was particularly true in the late 1999 dotcom boom when my venture was still a gestating idea. When we set out for the B-Series funding in early spring 2000, the economy was still booming, but we were expected to know considerably more about our technology, market, and capital needs. Between the March 23, 2000 meeting when the B-Series monies were committed and May 1 when the round closed, the macro economy had turned, and already the questioning became more intense.

Knowing *who is the audience* is fundamental to most good writing. In business-plan writing this is complemented by the question of *what decisions are being made*. Early in the venture, the audience is almost always funding sources. But even funding sources differ greatly in the writing they expect. Providing a business plan to support a National Institute of Standards and Technology (NIST) grant calls for different writing than for supporting an appeal to a venture capital fund. The aim of the former is to find platforms technologies that kick-start businesses broadly, while the latter is focused on growth shareholder value and a viable exit strategy.

Let's consider the task of writing the SDC business plan for Series-B funding. In today's just-recovering economy, this would be more like the document developed for the first round of venture funding. So the time frame for venture history is soon after forming the legal framework for the venture, with seed money in hand and about a half-dozen employees. Before we began writing, we sketched out the company in terms of the 11 dimensions Slywotzky[76] uses to characterize companies and the policy decisions they face. This sketch is presented below:

1. Fundamental Assumptions: *What are the fundamental assumptions behind the choice of this innovation in this particular market? How are customers changing? What are customers' priorities? What are the profit drivers for the business?* For SDC use of segment-based learning in technology-enabled marketing there were three fundamental assumptions: 1. Web businesses have been pushing for customer share. Now (2000) is the time to start pushing for performance out of their customer base, 2. Up-selling and cross selling from corporate databases are on the rise; and 3. Businesses spend most where the marginal productivity of capital is highest.
2. Customer Selection: *Which customers do I want to serve? Which ones will drive value growth?* SDC focused on corporate customers with SQL-based customer databases.
3. Scope: *What products and/or services do I want to sell? Which support activities do I want to perform in-house? Which ones do I want to subcontract or outsource?* SDC chose a bowling-alley strategy: Pick one business segment at a time and roll out services/software. Build knowledge across database platforms, and then build knowledge across industry segments.
4. Differentiation: *What is my basis for differentiation, my unique value proposition?* SDC believed it had a better mousetrap: faster, more scalable, and able to translate rules to action. *Why should the customer want to buy from me?* Clients don't have to move their strategic customer data outside the protection of their corporate firewalls: Incremental learning and action can be remote. *Who are my key competitors?* NetPerceptions for collaborative filters, BroadVision for manual rules, E.piphany

---

[76]Slywotzky, Adrian J. (1996), *Value Migration: How to Think Several Moves Ahead of the Competition.* Boston: Harvard Business School Press.

for legacy database integration and some datamining techniques, DataSage for large-scale analytical capability, and Personify for report generation. *How convincing is my differentiation relative to theirs?* There are a lot of inflated claims in this arena. Most customers are not convinced.

5. Value Recapture: *How does the customer pay for the utility I provide?* Value-based pricing. *How are my shareholders compensated for the value I create for the customer?* Investment rounds are in preferred shares – annual interest premium and capital appreciation at the time of a liquidity event.

6. Purchasing System: *How do I buy? Transactional or long-term relationship? Antagonistic or partnership?* We would probably have to purchase early on, until credit ratings were strong enough to get good leasing terms.

7. Manufacturing/Operating System: *How much do I manufacture versus subcontract? Are my manufacturing/service delivery economics based primarily on fixed or variable costs? Do I need state-of-the-art process technology?* SDC was in the digital economy. Almost all development costs are sunk, and variable costs are small in comparison.

8. Capital Intensity: *Do I choose a capital-intensive, high-fixed-cost operating system, or a less capital-intensive, flexible approach?* SDC would require $3-5 million in working capital over 18 months.

9. R&D/Product Development System: *Internal or outsourced? Focused on process or product? Focused on astute project selection? Speed of development?* SDC's strategy was to deploy across SQL platforms and segments, learn about segment problems to come up with "solutions," and move toward mining of Web sites as the "tornado" opportunity.

10. Organizational Configuration: *Centralized or decentralized? Pyramid or network? Functional, business, or matrix? Internal promotion or external hiring?* SDC had a Technology Office (Giuffrida), Business Development and Marketing (Garrett), Operations (VanArsdale), and CEO (Cooper). The organization was relatively flat, with the most emphasis on technology development.

11. Go-to-Market Mechanism: *Direct sales force? Low-cost distribution? Account management? Licensing? Hybrid?* Direct sales force is most likely for Gold client development. Silver clients could be found by Web self-identification and trade shows.

The business plan begins with the Executive Summary presented in Chapter 3. Only those who are sufficiently interested will continue reading the text of the business plan. A simple Table of Contents can communicate the basic outline of the business plan.

```
Executive Summary                        1
Table of Contents                        2
About Strategic Decision Corp.           3
About PersonalClerk                      3
    Benefits                             4
    Technology                           5
    Client Service & Support             6
    Target Clients                       7
Industry Overview and Competition        8
    Industry Background                  8
    Competing Technologies               8
    Competitors                          9
Financial                               10
    Pricing                             10
    Financial Forecasts                 11
    Planned Forecast                    12
    Limited Forecast                    13
Appendix                                14
    Management Team                     14
    Engagement Structure                17
    PersonalClerk Components            17
```

In the company summary (About Strategic Decision Corp.), you'll find echoes of the key terms the patent attorneys liked, but in the applied business context of up-selling and cross-selling. The idea is to convey the business benefit that we uniquely deliver. The company's commitment to providing a *whole-product solution* comes through clearly. The faculty connection provides an edge, particularly prior to having an installed base for the innovation, over the marketing messages of competitors. Faculty entrepreneurs need to become comfortable seeing their names and reputations used in this manner. If the product or service being offered by the new venture is not consonant with what you stand for in your university role, you are probably making a mistake by associating your name with the venture.

About Strategic Decision Corp.

> The Strategic Decision Corp. merges an innovative technological infrastructure with leading-edge customer analysis to provide secure personalized real-time marketing actions – up-selling, cross-selling, customer retention, and customer service. The company's technologies and services provide a remotely controlled, secure, resident, real-time datamining agent that improves the efficiency and effectiveness of marketing campaigns. The company's engagement teams provide whole-product solutions to dotcom companies that require strategic value from their customer databases, yet do not wish to compromise consumer privacy.
>
> Strategic Decision Corp.'s core product suite, **PersonalClerk**, integrates top-notch applications in computer science, marketing science, and management science. PersonalClerk never forgets customers' names or demographics, knows what they have bought in the past, and uses that knowledge to produce real-time, highly reliable offers of goods and services tailored to their preferences and circumstance.
>
> Professor Lee Cooper of the Anderson School at UCLA founded Strategic Decision Corp. in 1999 to leverage the tools and concepts he developed over 30 years of teaching, researching, and publishing about marketing science. The company's other management and technology team members are experienced marketing, technology, and consulting professionals with extensive knowledge in marketing strategy, database mining and analysis, expert systems, network implementation and security, and project management and implementation. In addition, the company has a distinguished panel of academic advisors contributing a level of intellectual capital to the company's development rarely enjoyed by other companies operating today.

PersonalClerk is the new concept, and obviously requires clear description. Juxtaposing the *new* with justification from recognized sources increases credibility. And thus, the quotes from Jupiter Communications and Inter@ctive Week play an important role. This is obviously not the place for equations, or for black boxes that are best inscribed with "then a miracle happens." The writing itself has to evoke plausible images of a systematic approach that could be specified in technical detail if the audience were different. Published articles or issued patents on relevant technology enhance credibility.

> About PersonalClerk
> PersonalClerk merges several proprietary tools for improved e-commerce marketing and customer service. At the heart of PersonalClerk is SCOPE, the **S**egment and **C**ustomer **O**riented **P**reference **E**ngine. SCOPE is a package of intelligent tools, uniting Strategic Decision Corp.'s market segmentation methodology with patent-pending technology for customer and transaction analysis.
>
> ***While close to 80 percent of Web site executives maintain that they monitor the behavior of repeat visitors, a much smaller percentage are in a position to leverage consumer data for effective targeting.***
> -Jupiter Communications "Proactive Personalization: Learning to Swim, Not Drown in Consumer Data," August 1999
>
> SCOPE builds and updates customer profiles incrementally with its unique, real-time datamining technology.

You need to clearly state the *core benefit proposition*. The benefits need to be those that a client would recognize, *increased offer effectiveness* and *security* in this case, rather than arcane advantages only meaningful to technological enthusiasts. Remember, this is the time for *crossing the chasm*. The client is the business decision maker, and the audience for this business plan wants to know how you will speak to the client. Benefits need to be translated into expected results. The results you promise will likely turn into the yardstick against which your value is measured.

> Benefits
> PersonalClerk offers the following key client benefits in one unique package:
>
> Improved Offer Effectiveness
> o **Personalization**—By profiling individual visitors to the Web site, PersonalClerk remembers customers' names and preferences, recognizing them upon return.
>
> o **Segmenting and Targeting**—Utilizing ZipSegments, Strategic Decision Corp.'s proprietary demographic

segmentation product, PersonalClerk offers clients the ability to market products that are consistent with customer demographics. In addition, PersonalClerk allows clients to recommend products to specific customer target groups. This is difficult with competitive artificial intelligence (AI) or "black-box" solutions, where customer segments and attributes are ambiguous.

- **Purchase event feedback**—PersonalClerk continuously gathers customer feedback used by SCOPE to update and refine future product suggestions.
- **Clickstream**—PersonalClerk can actually target offers to current and potential customers based on their pattern of movement through the client Web site, even before any financial transaction.

The result: increased sales through higher rates of conversion from Web browsing to purchase, larger order sizes, improved sales margin, and increased customer satisfaction.

Datamining is struggling to move toward real-time access to historical sales and customer data in order to tailor a more personalized approach to customers... [Current] efforts fall short of the goal: personalized, one-to-one, real-time marketing.
- Inter@ctive Week, "Data Miners Dig for Pay Dirt," Charles Babcock, December 5, 1999

Security

PersonalClerk respects all firewalls. Figure 6.2 illustrates PersonalClerk's location within the client's internal network. SCOPE analyzes and segments the customer data without absorbing, transferring, or removing customer records from the client's site.

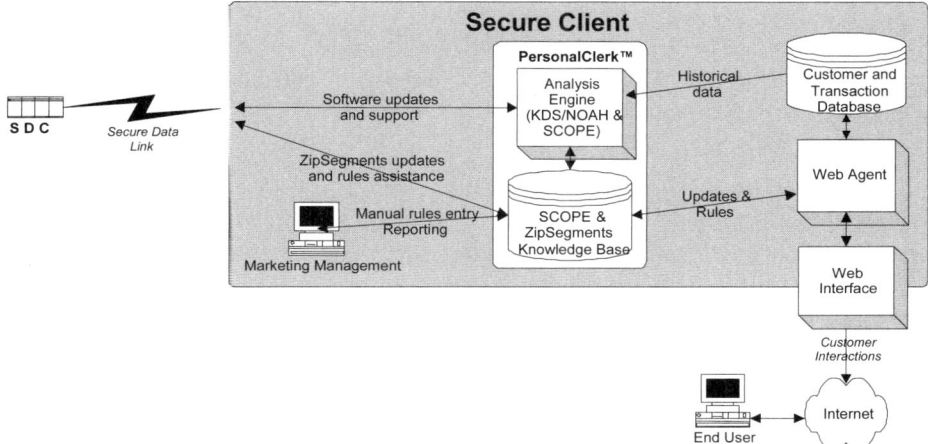

Figure 6.2. PersonalClerk's Communication with the Client Network

PersonalClerk's unique and secure design complies with all industry and
regulatory privacy standards. By employing PersonalClerk, clients can capitalize on their data to better serve customers, while minimizing exposure to consumer privacy concerns. In support of this, Strategic Decision Corp. is seeking a privacy certification from an appropriate independent supervisory body on the Internet.

Following the core benefit proposition comes another opportunity to drill down a little deeper into how those benefits are delivered (i.e., the components of the technology and the structure of client services and support) and who are the target clients.

Technology

PersonalClerk's seamless technology provides a comprehensive and unique experience to the end-user. PersonalClerk's analytical muscle is powered by these internally engineered, proprietary software-based technologies:

- **KDS/NOAH**—Strategic Decision Corp.'s optimized rules generating dataminers, determining the product suggestions directed to each segment and customer and identifying cross-selling opportunities within and between product categories

- **ZipSegments**—Strategic Decision Corp.'s exclusive segmentation scheme, classifying U.S. ZIP codes into 68 consumer segments based on the demographic data in the United States Census

- **SCOPE**—Strategic Decision Corp.'s Segment and Customer Oriented Preference Engine working in combination with KDS/NOAH and ZipSegments to provide automated up-selling and cross-selling recommendations

These four components of PersonalClerk work together to deliver a customer relationship management system that is self-educating, current, and accurate as to customers' and prospects' demands. Figure 6.3 outlines how PersonalClerk utilizes data from various sources to provide recommendations and targeted offers through SCOPE.

Figure 6.3. PersonalClerk Utilizes a Variety of Data Sources to Provide Real-time Marketing Messages

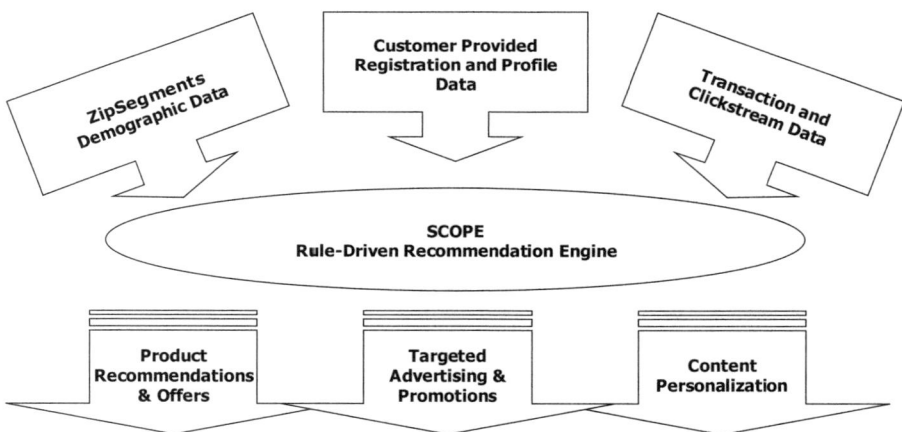

Client Service & Support
Strategic Decision Corp. provides a complete package of services enabling clients to realize the full value of the PersonalClerk product suite:

- **Implementation**—Engagement teams provide clients with implementation services to install the product and build the appropriate links to their respective databases and Web sites.

- **Ongoing Support**—The support organization assists all licensed clients with the continuous use of the PersonalClerk software. The suite of services is customized according to client needs. Support packages range from simple troubleshooting to full, ongoing marketing program assistance, database management, and rules-generation outsourcing.

- **Updates**—Following implementation, the company routinely improves and updates the ZipSegments database and the PersonalClerk software components. Strategic Decision Corp. remotely monitors the agent and submits new rules, improving the quality of the software and rules set.

Engagements are structured to meet individual client needs, with varying degrees of client staff participation and data integration. Strategic Decision Corp. will also offer consulting services, leveraging our rich academic and analytical resources, for clients requiring additional assistance.

Target Clients
The PersonalClerk technology is equipped for clients who maintain and build customer databases. Target clients include:

- Individual electronic retailers—improving sales productivity

- Retail aggregators or "e-malls"—leveraging customer data across product categories to cross-sell

- Service providers—marketing value-added services and products to their customers

- Traditional businesses—capitalizing on large customer databases to market additional services

- Content providers—offering advertisers improved targeting

Strategic Decision Corp.'s initial focus is distributing PersonalClerk's customer data analysis and real-time recommendation technology to the Internet retail marketplace. "E-retailers" provide the greatest opportunity for rapid implementation, allowing the company to demonstrate PersonalClerk's strengths while at the same time participating in this fast growing market. By the middle of year one, the company will pursue targeted advertising campaigns and the online retail mall marketplace. PersonalClerk's technology is perfectly suited to assist online advertisers in conducting segment-specific marketing campaigns, dramatically increasing returns on advertising dollars. Online malls provide the opportunity to serve a number of retailers in a common environment, potentially leveraging customer information across the mall. At the same time, customer information is retained by the mall and not shared directly with member stores, protecting privacy.

New ventures, particularly those stemming from academic roots, establish a lot of the credibility through their understanding of market risk. This may not be the most important area in general, but is often the weakness of ivory-tower ventures that potential funding sources pick on.

Industry Overview and Competition

Industry Background

Over the past five years, the Internet has developed into a powerful sales channel for retailers. To date, these "e-retailers" have mainly focused their resources on e-commerce-enabling technologies and advertising to acquire new customers. They have also established methods to track customer behavior online, and in some cases, deliver targeted, but rudimentary, marketing messages to individual customers.

According to Jupiter Communications, a leading Internet research firm, e-retail strategy will shift by the end of 2001 from a focus on customer acquisition to customer retention. To meet this challenge, e-retailers will demand more sophisticated Web marketing tools that will allow them to better understand their customers and provide them with personalized online experiences.

Addressing this need is broadly defined as "personalization." Personalization includes capturing data on individual customers and using it to deliver targeted

> marketing recommendations and messages. The process requires building a flexible data management infrastructure that accommodates a variety of consumer data and analytical techniques.
> *The next wave of spending will focus on targeted solutions that capitalize on the powerful information sharing and distribution capabilities of the Web, enabling personalized Web experiences for consumers, including real-time, targeted marketing and sales messaging. Strategic Decision Corp. will fulfill this need with its products and services.*

You will need to discuss competitive technologies and companies in a way that leads to a figure such as Figure 3. We chose to develop this figure in terms of the underlying competitive technologies. This was partly due to the imprecise descriptions on some competitors' websites and the continuous inflow and outflow of companies in the rapidly changing market. The discussion of the competing technologies is the first chance to set up the frame of reference for comparison. Cooper (2003) deals with some of the substance of this approach versus the alternatives. Here I emphasize the function of Figure 6.4. The columns represent the desired attributes of business solutions in this domain, and the rows reflect the competitive technologies. The right chart has your offering as the only one with checks in all columns, the comprehensive solution. It is not as simple as just picking the features unique to your offering. If the columns are not perceived as the important attributes, the reaction could be dismissive. If you omit alternative technologies so you look better, you will be viewed as naïve. You don't need to beat the competition on every point, just be the comprehensive alternative with no clear inferiority.

> Competing Technologies
> Four analytical technologies, discussed below, are currently in use or in development to address need for personalization. Each technique draws from different data sources, such as user platform/domain, click-stream, site registration, search query inputs, and stated preferences. Strategic Decision Corp. believes, however, that transaction data yield the richest consumer data, enabling a site to capture verifiable demographics that are crosschecked with purchase history profiles.
> The four competing analytical technologies in personalization are currently:

- ***Clustering***—Uses statistical techniques to group site visitors with similar characteristics into segments
- ***Profiling***—Characterizes individual consumers based on their interaction with Web site elements
- ***Collaborative Filtering***—Uses algorithmic techniques to infer preferences based on similar behavior from others
- ***Artificial Intelligence or "AI"***—A range of technologies, including natural language processing, expert systems, and neural networks

According to a recent Jupiter Executive Survey, "…interest has migrated to the application of more transparent systems used to model consumer data and target users. Jupiter Executive Survey findings indicate a growing interest in tools that model online behavior, and in datamining tools (clustering techniques), while interest in advanced AI solutions is waning."[77] Other than clustering, the technologies listed above are "black box" solutions, referring to the inability of marketing managers to understand or manage the actions of the technologies.

PersonalClerk uses a combination of clustering and profiling techniques to model customer behavior from actual purchase transactions. Furthermore, it is the only application that offers a real-time messaging component, a proprietary segmentation scheme (ZipSegments) and a patented scoring methodology (SCOPE) to deliver highly targeted marketing recommendations and messages.

Competitors
Companies offering customer analysis tools for the Web include E.piphany, DataSage, Broadbase, Personify, and NetPerceptions. A key difference between most of these companies and Strategic Decision Corp. is the segmentation methodology that supports management decision-making. Most competitors use various collaborative filtering and neural-network technologies in which consumer attributes are difficult to discern, while interest is moving toward real-time clustering techniques.

---

[77]Jupiter Communications "Proactive Personalization: Learning to Swim, Not Drown in Consumer Data," August 1999.

Figure 6.4 highlights the advantages of PersonalClerk versus competitive offerings.

Figure 6.4. Competitive Customer Analysis Technologies.

	Clustering of consumers to support targeted marketing	Analyzes behavior and generates rules in real-time	Supports management decision-making (transparency)	Secure environment
PersonalClerk	✓ Full demographic and preference based clustering	✓ Incremental mining to update rules immediately	✓ Supports management learning and intervention in rules and preferences	✓ Customer data remains within site infrastructure
Collaborative filtering technologies	Limited "Buying groups" not linked to demographics or broader behavioral data	No Batch update can require hours or days	No Product linkages provide little insight or opportunity for intervention	Depends on implementation
Neural-net / "AI" technologies	Limited "Clusters" have little relevance outside of AI system	No Batch update can require hours or days	No Data is caught in "black-box" without meaningful management learning	Depends on implementation

In addition, there are a number of vendors with highly developed solutions in the front-end and database management markets, including Microsoft and Oracle. Traditional enterprise management and sales force automation vendors such as SAP and Siebel also may wish to provide personalization products to complete their package of products. Nevertheless, these companies are potential partners rather than direct competitors. Strategic Decision Corp. maintains significant, sustainable advantages:

- These companies lack strong real-time data analysis expertise and technology, requiring capabilities significantly different from their core competencies.

- These companies' current products necessitate a large up-front software investment by clients rather than ongoing analytical support; data analysis solutions require a much higher degree of hands-on assistance to be used effectively. It is unlikely that these firms will migrate to a significantly different revenue and support model.

- Strategic Decision Corp. uniquely has patents pending on its leading-edge technologies, giving the company a significant advantage over all potential competitors.

Strategic Decision Corp. is the only company delivering targeted marketing messages in real time by combining highly secure datamining with real-world segmentation, personalization, real-time learning, and purchase event feedback.

Value recapture, the revenue model, is an essential part of a business plan supporting funding decisions. Despite the inflated memories of an era of easier money, plans that focused on revenue and ignored costs were the exception, never the rule. What was more usual were plans that based revenue on 1 percent of large markets without suggesting how that 1 percent would be obtained. This hubris was then confounded by compounding that 1 percent by the growth rate for the Internet sector of that larger market. Most of these "hockey stick" models never saw the puck go in the goal. Revenue from these unrealistic models so far outstripped conceivable costs that great profits were expected. The VCs in those cases did not ask enough tough questions. Revenue models based on "monetizing eyeballs" were clearly dependent on an Internet advertising market, which, even at that time, I felt had to be tied to transactions to be viable. "Selling customer data" was another mysterious source of revenue cited by data-intensive companies that I did not believe would pass muster with VCs. Our pricing was based on getting customers and providing value for those customers. The amount of our charges under either of the schemes offered was based on our sense of the value provided. The bowling-alley strategy (Moore 1995) clearly says that value-based pricing is appropriate at this stage of venture development. The two pricing models reflected our uncertainty over the acceptability of value pricing in this market. We wanted to align our incentives with those of our clients, but at this point had no evidence of the acceptability of this in our market.

Financial

Pricing
Strategic Decision Corp. will price its services in two categories:
1. Up-front charges—These include the costs of implementing PersonalClerk, and vary widely depending on the complexity of the client environment and the degree of assistance required to prepare their database(s). At a minimum, these charges cover all direct time and expenses associated with the installation.
2. Ongoing charges—These cover the continued use and support of PersonalClerk and updates to ZipSegments. Depending on the client situation, Strategic Decision Corp. will offer the following options:

- Flat licensing, based on the complexity and size of the clients' data set
- Percentage of assisted sales, where Strategic Decision Corp.'s earnings would result from increases in our clients' sales

Each pricing method is appropriate for different client situations, and the company's sales force will employ them selectively to maximize growth and profitability.

	Up-front charges	Ongoing charges
Flat pricing	$100 - $200K	$100 - $250K over four years
Percentage of sales	$50 - $150K	.3% - .6% of assisted revenue = ~$200 - $550K over four years

Add-on services
Many clients will further employ Strategic Decision Corp.'s rich marketing and customer database skill-set to enhance their marketing programs. Such services could include assistance with marketing objectives or technical services emerging from those normally provided during implementation or support of the PersonalClerk product. The basic method of calculating charges for add-on services is a time-and-materials basis, although other methods will be considered depending on the client's situation. Conservatively, the company does not capture revenue streams from these services in the financial model.

The dual cash-flow forecasts were our nod toward best-case and worst-case scenarios. Steve Mayer tells the story during his funding search for Digital F/X of being asked for a worst-case scenario. As with the one we presented, he talked about falling a little short on this goal and that one. The veteran venture capitalist looked at him and said nobody gives a real worst case: that the product will completely fail, and we'll end up bankrupt and being sued for the next 15 years. The projected cash flows presented below merely signal that somebody on the team has an MBA and a spreadsheet underlies our forecasts. If someone asks how many clients we are anticipating, we have an answer. Ask how many people serve each client, and we have an answer. How long does this round of cash last? We have an answer.

Financial Forecasts

Two Strategic Decision Corp. financial forecasts have been prepared:

The "planned" forecast represents management's current estimate of revenue growth, averaging over 300% annually, with related expenses. Under this forecast, total fees will reach more than $900,000 per month by the end of year one and approximately $5 million per month by the end of year two. With these revenues, monthly income will be positive by March of 2001. These projections are blended to include an equal percentage of customers using the flat licensing option and the value-based ongoing fee option. Value-based fees should generate higher long-run revenues than flat fees, but depend heavily on the actual growth in revenues of the clients.

The "limited" forecast represents an alternative approach with lower-than-expected client revenues, yielding approximately 250% average annual growth, with appropriate expenses. Under this scenario, total fees are estimated to reach $650,000 per month by the end of year one and approximately $2.5 million per month by the end of year two. With these revenues, monthly income will become positive by July of 2001. Under this scenario, projections include only 25% of clients utilizing the value-pricing model, with the remainder using the flat price licensing option, with total pricing approximately 18% lower.

These prices and growth rates compare favorably with published costs for comparable technologies' fee and growth data. Competitive research suggests that recommendation technologies generally cost $150,000 for implementation, with comparable ongoing costs. PersonalClerk also provides customer knowledge management capabilities. In addition, Jupiter Communications cites implementation budgets of $1M to $4M for customer relationship management, or CRM, software.[78] Likewise, the Gartner Group estimates that typical outsourced CRM solutions cost approximately $1-5 million annually in licensing and operating costs.[79] CRM is the broader industry segment, in which our software participates, and these total budgets include internal implementation costs and the cost of database management tools, so our products and services would represent a portion of these budgets.

Growth estimates compare favorably with various industry and analyst estimates. International Data Corporation estimates that the amount of Internet commerce worldwide will increase from $32 billion in 1998 to more than $400 billion in 2002. PersonalClerk and its underlying technology directly affect the competitiveness of retailers and advertisers conducting Internet commerce. We expect to grow both by expanding the percentage of these retailers and advertisers utilizing our products and through the continued expansion of this market.

---

[78]Proactive Personalization, Jupiter Communications, December 1999.
[79]CRM ASP Opportunities, Dataquest, August 23, 1999.

Table 6.1. Financial Forecast Detail – Planned Forecast

Pro-form Projected Cash Flows

	2000	2001	2002	2003
Starting Cash	$ -	$ 22,661	$ 108,840	$ 157,308
Investment	$ 6,236	$ -	$ -	$ -
Q1 Revenues	$ -	$ 2,992	$ 15,488	$ 41,607
Q1 Expense	$ 810	$ 4,003	$ 9,982	$ 13,919
Q1 Inv in Equipment	$ 98	$ -	$ -	$ -
Q1 Income	$ (908)	$ (1,011)	$ 5,506	$ 27,688
Taxes	$ -	$ (303)	$ 1,652	$ 8,306
Ending Cash	$ 5,328	$ 21,954	$ 112,694	$ 176,689
Starting Cash	$ 5,328	$ 21,954	$ 112,694	$ 176,689
Investment	$ -	$ -	$ -	$ -
Q2 Revenues	$ 476	$ 5,145	$ 21,320	$ 49,625
Q2 Expense	$ 1,383	$ 4,240	$ 10,286	$ 14,059
Q2 Inv in Equipment	$ 182	$ 13	$ 4	$ 1
Q2 Income	$ (1,090)	$ 891	$ 11,029	$ 35,566
Taxes	$ (327)	$ 267	$ 3,309	$ 10,670
Ending Cash	$ 4,565	$ 22,577	$ 120,414	$ 201,585
Starting Cash	$ 4,565	$ 22,577	$ 120,414	$ 201,585
Investment	$ -	$ 80,000	$ -	$ -
Q3 Revenues	$ 1,524	$ 8,690	$ 30,745	$ 63,612
Q3 Expense	$ 2,935	$ 5,797	$ 11,269	$ 15,230
Q3 Inv in Equipment	$ 273	$ 79	$ 12	$ 11
Q3 Income	$ (1,685)	$ 2,815	$ 19,464	$ 48,371
Taxes	$ (505)	$ 844	$ 5,839	$ 14,511
Ending Cash	$ 3,386	$ 104,547	$ 134,039	$ 235,445
Starting Cash	$ 3,386	$ 104,547	$ 134,039	$ 235,445
Investment	$ 20,000	0	0	0
Q4 Revenues	$ 2,801	$ 14,874	$ 46,651	$ 82,249
Q4 Expense	$ 3,759	$ 8,639	$ 13,379	$ 17,159
Q4 Inv in Equipment	$ 77	$ 103	$ 30	$ 13
Q4 Income	$ (1,035)	$ 6,132	$ 33,242	$ 65,078
Taxes	$ (310)	$ 1,840	$ 9,972	$ 19,523
Ending Cash	$ 22,661	$ 108,840	$ 157,308	$ 280,999
Annual Revenue	$ 4,801	$ 31,700	$ 114,204	$ 237,093
Annual Expense	$ 8,888	$ 22,679	$ 44,917	$ 60,367
Annual Income	$ (4,717)	$ 8,827	$ 69,240	$ 176,702
Cumulative Income	$ (4,717)	$ 4,109	$ 73,349	$ 250,051

Table 6.2. Financial Forecast Detail – Limited Forecast

Pro-form Projected Cash Flows

	2000	2001	2002	2003
Starting Cash	$ -	$ 22,410	$ 103,854	$ 122,327
Investment	$ 6,236	$ -	$ -	$ -
Q1 Revenues	$ -	$ 2,529	$ 8,114	$ 17,713
Q1 Expense	$ 810	$ 3,979	$ 5,842	$ 6,884
Q1 Inv in Equipment	$ 98	$ -	$ -	$ -
Q1 Income	$ (908)	$ (1,450)	$ 2,272	$ 10,829
Taxes	$ -	$ (435)	$ 682	$ 3,249
Ending Cash	$ 5,328	$ 21,396	$ 105,444	$ 129,907
Starting Cash	$ 5,328	$ 21,396	$ 105,444	$ 129,907
Investment	$ -	$ -	$ -	$ -
Q2 Revenues	$ 431	$ 3,534	$ 10,404	$ 22,606
Q2 Expense	$ 1,383	$ 4,006	$ 5,906	$ 7,238
Q2 Inv in Equipment	$ 182	$ 1	$ 1	$ 3
Q2 Income	$ (1,134)	$ (474)	$ 4,497	$ 15,365
Taxes	$ (340)	$ (142)	$ 1,349	$ 4,610
Ending Cash	$ 4,534	$ 21,064	$ 108,592	$ 140,663
Starting Cash	$ 4,534	$ 21,064	$ 108,592	$ 140,663
Investment	$ -	$ 80,000	$ -	$ -
Q3 Revenues	$ 1,384	$ 5,420	$ 13,728	$ 29,481
Q3 Expense	$ 2,892	$ 4,204	$ 6,250	$ 8,033
Q3 Inv in Equipment	$ 259	$ 11	$ 4	$ 6
Q3 Income	$ (1,767)	$ 1,205	$ 7,475	$ 21,441
Taxes	$ (530)	$ 362	$ 2,242	$ 6,432
Ending Cash	$ 3,297	$ 101,908	$ 113,824	$ 155,672
Starting Cash	$ 3,297	$ 101,908	$ 113,824	$ 155,672
Investment	$ 20,000	0	0	0
Q4 Revenues	$ 2,556	$ 7,926	$ 18,899	$ 38,634
Q4 Expense	$ 3,739	$ 5,112	$ 6,746	$ 9,424
Q4 Inv in Equipment	$ 84	$ 35	$ 6	$ 12
Q4 Income	$ (1,267)	$ 2,780	$ 12,146	$ 29,198
Taxes	$ (380)	$ 834	$ 3,644	$ 8,759
Ending Cash	$ 22,410	$ 103,854	$ 122,327	$ 176,110
Annual Revenue	$ 4,372	$ 19,409	$ 51,145	$ 108,433
Annual Expense	$ 8,825	$ 17,301	$ 24,744	$ 31,579
Annual Income	$ (5,076)	$ 2,062	$ 26,390	$ 76,834
Cumulative Income	$ (5,076)	$ (3,014)	$ 23,376	$ 100,210

The forecasts we presented were far different from what actually happened. When that happens, you must ask the three standard questions (Cooper and Nakanishi 1988, p.15): Were the forecasts of industry sales off? Were the forecasts of our market share off? and, Were the marketing activities carried out as planned? The industry (e-retailing) grew modestly in 2000, then grew robustly year after year. But our share of that sector was off most fundamentally because the new CEO moved the company out of e-retailing and into Internet advertising on the eve of a crash in the advertising market.

The appendix spelled out details that responded to other potential questions. The organization chart can be full of TBDs (to be determined), as long as the key roles are filled and an interim reporting structure is indicated. I preferred being listed as simply chairman of the board, but I was the chief executive officer, and not listing who has that authority is a mistake. Not having a CFO in a 10-person company is fine as long as you lay out who is responsible for that function.

Appendix
Management Team
We have organized Strategic Decision Corp. as follows:

Figure 6.5. Organization Chart.

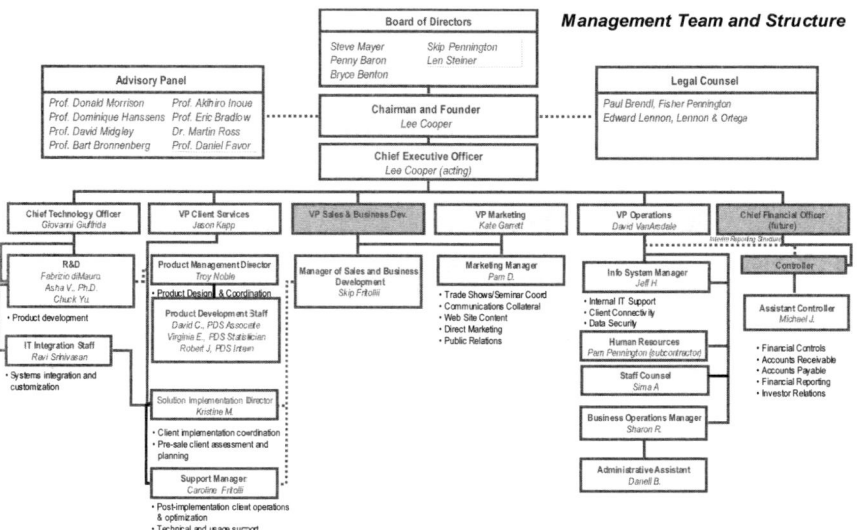

This was followed by a paragraph on each person on the senior management team, board of directors, and academic advisor panel. We noted that Paul Brendl, partner, Fisher Pennington, LLP; and Edward Lennon, founding partner, Lennon & Ortega, LLP, represented us. We also listed our major investors at that point: Bud Pennington, founding partner, Fisher Pennington, LLP; Fred Hart, Hart Media, Inc.; Len Steiner, chairman, Steiner & Steiner, Inc.; Jay Hillis, CFO, Hart Media, Inc.; and additional Fisher Pennington attorneys.

We also wanted potential investors to know we had thought through how we would interact with clients. Jason Kapp's IT consulting experience led to an engagement structure in which our internal structure mirrored the personnel on the client side so as to provide clear lines of communication between organizations and clear lines of authority within organizations:

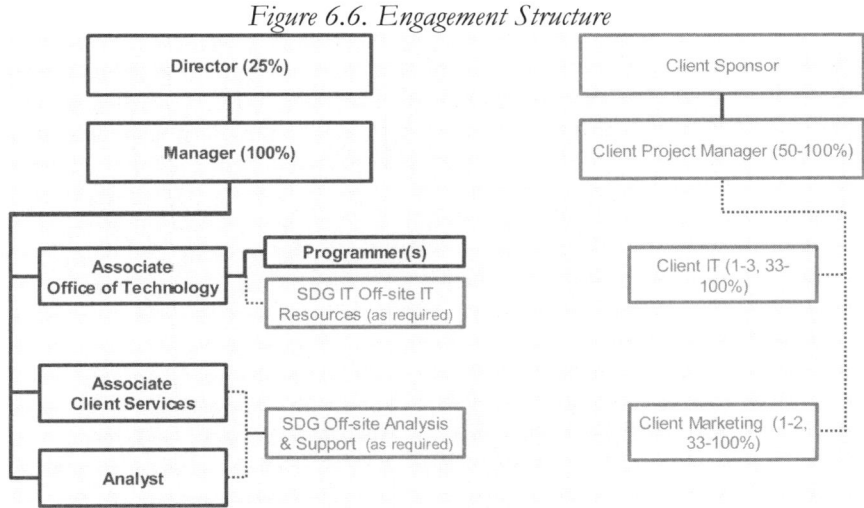

Figure 6.6. Engagement Structure

The rest of the appendix provided a little more detail on components mentioned in earlier parts of the business plan.

PersonalClerk Components

KDS
KDS is a rule-generating dataminer implemented as a superset of SQL. It is the technology behind the forthcoming *Management Science* article "Turning

Datamining into a Management Science Tool." This is the first datamining article to pass the rigorous peer-review process in the flagship journal of the Institute for Operations Research and Management Science (INFORMS). KDS organizes knowledge into a hierarchical structure, enabling rapid identification of all rules that apply to a new "case." We have optimized KDS for databases with a large number of records and a small number of independent variables, each of which has a large number of levels or features (i.e., number of levels of a nominally valued independent variable). We proved it is "best of class" when mining a 1.3 million record database with more than 4,500 features.

NOAH

NOAH is our datamining algorithm, optimized for cross-selling applications on customer databases. We developed NOAH utilizing a customer database with more than 40,000 customers, where each customer has 221 attributes for 53,000 "features." This database was also missing a large amount of data, typifying direct marketing databases. The paper describing this software, "NOAH: An Algorithm for Mining Classification Rules from Datasets with Large Attribute Space," was accepted for the EDBT 2000: Proceedings of the Conference on Extending Database Technology, March 27-31, 2000, in Konstanz, Germany.

ZipSegments

ZipSegments is Strategic Decision Corp.'s exclusive geo-demographic segmentation scheme, driven by the connection of ZIP codes to data from the U.S. Census. In this way, clients have access to detailed geo-demographic and socioeconomic data about populations within each ZIP code. All clients will be encouraged to use ZIP codes as an integral part of customer identification. Once the ZIP code is obtained, PersonalClerk can make better offers and recommendations based on this knowledge.

SCOPE

SCOPE is a **S**egment and **C**ustomer **O**riented **P**reference **E**ngine. As customers browse a Web site, events occur. Each click of the mouse is an event that, when collected, provides data about customers' product and service interests. The Web site will also ask the customer to respond to questions. The response is also an event that is collected. SCOPE monitors these events and aligns them against a pre-determined segmentation pattern. For each segment, SCOPE accumulates the selections of segment members. The software records the popularity of each option within each segment, noting the popularity of selection without recommendation and the popularity of the selection when recommended. The software then recommends the most popular offers to the customer, after filtering for those options already selected in the customer's selection history. If the customer selects the recommendation, the popularity of that selection increases in the hierarchy of offers. If the site presents the customer with the recommendation and the customer declines the offer, the popularity of that offer decreases.

6.5 Due Diligence on the Business Plan

As indicated earlier in this chapter, you must expect due-diligence inquiries any time a financial decision is being made. An Internet infrastructure expert interviewed Giovanni and me extensively prior to the seed funding (Series A). A top advisor to the FCC carefully reviewed our initial security design before the seed round. Prior to the Series B funding, due diligence was conducted privately by AT Kearney before their offer of $1 million (half in in-kind services) in that round, which we did not accept. The private client group at Merrill Lynch brought a PhD/CFA from Intelligent Technology Ventures to the first presentation on the Series B funding to conduct due diligence because of that fund's prior investment in and experience with similar companies. That was what precipitated the ultimate involvement of that venture fund in the B Round. But before the final decision, a JD/MBA from Intelligent Technology Ventures interviewed every employee in the company. Edward Lennon (Lennon & Ortega) solicited a due-diligence summary from the intellectual property attorney he brought to work with us. I did not know that due diligence was being performed when I discussed IP with the attorney, but the lesson is that due diligence is *always* being performed. Expect it. When it turned out to be acceptable, I was given a copy, which I reproduce below to introduce you to what is going on whether you know it or not.

Dear Edward:

You asked me for a high-level summary of our evaluation of the Strategic Decision Corp. business opportunity. We have identified and addressed six issues below. Our summary conclusion as to each issue is placed in italics at the end of each paragraph.

Value Proposition. The company proposes to provide real-time marketing recommendations to vendors based on historical buying activities of a customer and on results for offers to customers in the same market segment, all while allowing the customer to keep its own data behind a firewall. We think this will be a very attractive proposition for vendors. ***On balance, we believe the value proposition is sound.***

Technical Due Diligence. Since the company's products are not yet fully defined, we cannot address every technical issue. For example, at some point in the life of the company, we might want to know whether hardware configurations are adequate, etc. At a high level, the company's approaches appear to be well thought-out and to offer a good possibility of allowing the company to execute on a combination of products and services that deliver on the value proposition. Lee Cooper's position in the academic community and the academic quality of the rest of the team add credibility to this conclusion. However, it is too early to tell what the costs will be in delivering on the proposition to customers as the business expands. The technical side of the company's offerings will remain a somewhat unknown quantity until a later stage of development. In sum, ***they can probably execute on the value proposition, but at what cost?***

Scalability. The business model calls for some customization of the company's offerings to specific customers and for the provision of various services related to implementation of the company's algorithms. For this reason, significant expert human resources will be needed in order to expand the business. Given the tight market for marketing expertise, this could present an inherent limitation on growth. Therefore, ***expanding to scale could be a problem.***

Competition. As noted in the company's business plan, there are many other companies interested in data mining, in a variety of fields. Some of those companies, e.g., Oracle, Sybase, etc., are very well funded. Therefore, the likelihood of the company's establishing a dominant position in the data mining field appears to be small. A stronger case can be made for building market share in a niche area, such as customers requiring real-time recommendations coupled with high database security. However, even customers in this niche are likely to be offered a wide range of different products and services to assist them with on-line marketing. Some of those other products and services might be complementary, but it might be difficult to get the attention of the customer in a crowded field. On balance, *competition will be a significant risk.*

Intellectual Property. Our searches have not found other companies that have product offerings or patents claiming the company's entire value proposition, i.e., a remote, secure, real-time data mining agent that permits a customer's data to remain behind the firewall. Thus, we are optimistic that the company will be able to obtain patent protection of reasonable breadth for its core offering. However, there are major caveats on this point. First, because pending patent applications are held in secrecy, and because patenting of Internet business models is a fairly new development, we would probably not be able right now to find any problems that do exist. Moreover, the crowded nature of the field prevents us from examining all other product offerings, even the ones that are public. Finally, as noted in paragraph 4 above, patent protection in the area of the company's direct product and service offerings will not preclude competition from other offerings that promise the customer similar benefits through different mechanisms. Therefore, *the proprietary position is a plus, but with significant caveats.*

Human Resources. The team has strengths and weaknesses, most of which are different sides of the same coin. Lee Cooper and Giovanni Giuffrida bring significant credibility to the venture in terms of their intelligence and their expertise on the underlying algorithms. In addition, Lee's personality and his position at UCLA suggest that he will have continued access to talent in areas of need as the company scales up, and that he will be good at bringing that talent into the company. On the other hand, it is unclear, at least to us, what Lee's long-term operational role with the company will be. The company will no doubt face many day-to-day operational challenges, and the company's academic origins raise questions about whether it has the experience or lines of structure to address these challenges. **On balance, human resources appear to be a plus, but raise some concerns as well.**

As always, please let me know if we can provide any further assistance.

Best regards,

(signed by the attorney specializing in intellectual property.)

7. Strategic Maps

This chapter presents case examples of strategic marketing planning. It develops the background for strategic decisions concerning retaining the broad approach to technology-enabled marketing or narrowing the focus to Internet advertising. It is written for advanced management students and MBA-trained managers who wish to learn how modern approaches to market assessment and strategic marketing planning can be applied to a new venture such as Strategic Decision Corp.

7.1   Strategy as Comprehensive Problem Solving

Much of the literature in marketing strategy is devoted to normative speculations, derived from economic modeling, that go something like: If the world were composed of N firms involved in (pick one) duopolistic, oligopolistic, or pure competition with a (pick one) differentiated or undifferentiated product, then the optimal (pick one) pricing, advertising, or distribution policy would be (insert answer). We postulate the number of players, the methods of interaction, and the rules of the game, and derive the optimal behavior under these conditions. The evolutionary-modeling alternatives seed the initial generation of players and the rules of creation, exchange, and death, and observe the emergent behavior in agent-based simulations. Regardless of whether you start with top-down derivations or bottom-up simulations, the results are vignettes yielding tactical insights, rather than comprehensive assessments on which you can base strategic marketing plans.

On the qualitative side are many conceptual models useful for problem articulation, such as Slywotzky's business-design framework described in Section 6.4[80] or Porter's Five Forces: Supplier Power

---

[80]Slywotzky, Adrian J. (1996), *Value Migration: How to Think Several Moves Ahead of the Competition.* Boston: Harvard Business School Press.

194 | Midlife ~~Crisis~~ Startup

(supplier concentration, differentiation of inputs, impact of inputs on cost or differentiation, presence of substitute inputs, and threat of forward integration); Buyer Power (bargaining leverage, buyer volume, price sensitivity, threat of backward integration, product differentiation, buyer concentration vs. industry, substitutes available, and buyers' incentives); Threat of Substitutes (switching costs, buyer inclination to substitute, and relative price performance of substitutes); Barriers to Entry (absolute cost advantages, proprietary learning curve, access to inputs, government policy, economies of scale, capital requirements, switching costs, access to distribution, and expected retaliation); and Degree of Rivalry (exit barriers, industry concentration, industry growth, switching costs, and diversity of rivals).[81]

A tactic is a solution to a particular problem in a given context. A strategy is an approach to addressing simultaneously the entire interdependent set of problems. From my point of view, a relatively comprehensive articulation of the problems and their interdependencies constitutes a prerequisite for strategy formulation. The first time managers of a new venture face the need to spell out the issues, the list may be simpler than it ultimately becomes. It is important only to have a place to begin and a path to make it better. The litany of issues need to be interrelated into a what-influences-what map of the strategic terrain. The case in point illustrates how that mental map can be made dynamic – enabling top managers to address the what-if questions they need to operate in a turbulent world.

As indicated in Chapter 2, a turbulent environment gives rise to many issues. To begin, I use a standard classification of environmental forces: political, behavioral, economic, sociological, and technological categories. Issues in these areas may affect the broader infrastructure, they might concern the particular business ecosystem in which the firm operates, or the issues may be unique to the venture itself. While historically I've presented these as forming a grid of five environments by three points of view, I view them conceptually as forming a target or map, with the company in the center, surrounded by the business ecosystem, which is itself surrounded by the infrastructure (see Figure 7.1). The issues central to the company also affect the business ecosystem and the infrastructure less directly.

---

[81]Porter, Michael (1980), *Competitive Strategy*, New York: The Free Press. Porter, Michael (1985), *Competitive Advantage*, New York: The Free Press.

Analogously, the infrastructure issues, of course, have an impact on the business ecosystem and the company. The solution to infrastructure problems may require cooperation among members of the business ecosystem that is unlikely for problems directly within the ecosystem. The nesting of business ecosystem with infrastructure, and company with ecosystem, helps identify when common efforts are more or less likely to be available for problem solving.

Figure 7.1. Critical Issues Map.

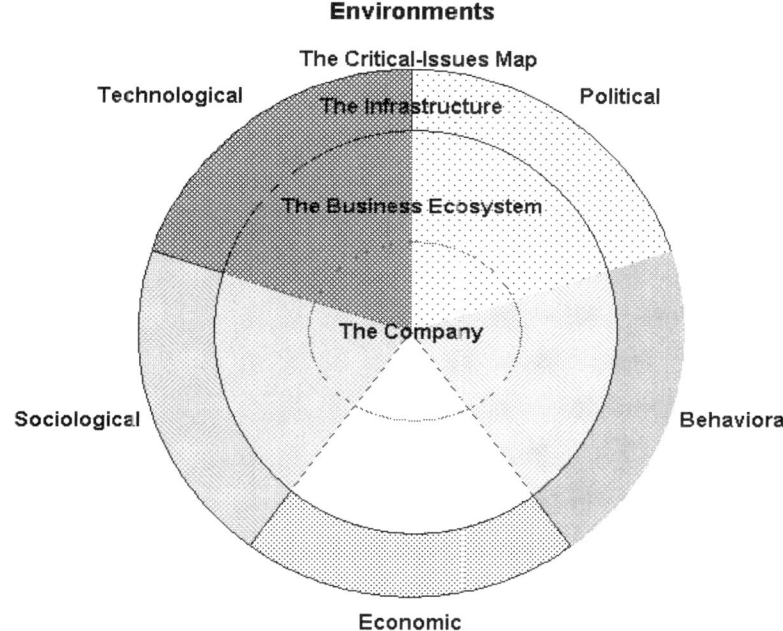

I begin this stage by developing a general description of the product or service, and noting the stakeholders in the outcomes. Then I list the issues prompted by each cell in the grid. This becomes an outline for short written statements about the issues. Sharing this document within a planning group is a good way to find out if there is agreement on the critical issues.[82]

---

[82]In what follow I have incorporated parts of the planning-project submitted by Ravi Narasimhan, Al Mamdani, Vijay Mididaddi, and Pak-yan (Eric) Liang for the Winter 2000 section of "Marketing Strategy in the Digital Economy."

For the case of Strategic Decision Corp., my planning focus centered on SDC's likelihood of success and the consequence of that success for valuation of the company. I am particularly concerned about the strategic decision between broadly pursuing technology-enabled marketing optimization (TEMO) as part of the knowledge-management sector and narrowing the focus to Internet advertising. I'll lay out the mental model I had during the late summer of 2000, and then run scenarios to show the impact of switching from technology-enabled marketing to Internet advertising as a specific focus.

In the Winter Quarter of 2000, the student team undertaking a planning exercise for SDC positioned the company in relation to four categories:

- Business Intelligence Products
- Relationship Management Products
- Data Warehousing Products
- Visualization and Reporting Products

Business Intelligence Products

1. Net Perceptions: This startup, based in Northern California, marketed a personalization software product based on collaborative filtering. Net Perceptions had a well-publicized IPO in May 1999, and has since made several moves to consolidate its position as the premier Web-retail personalization engine on the market. Rumored at the time to be spending 110% of revenue on marketing and sales, Net Perceptions was a very active competitor whom we saw walking out the door as we walked in to meet with potential clients. Net Perceptions offered what it called a real-time recommendation engine (although it was far from updating or learning in real time) that integrates with a retailer's e-commerce solution. Accordingly, it pursued several strategic alliances to *package* its product with commercial commerce engines; the largest deal up to the time of this planning effort was with ATG's Dynamo Product Suite of E-commerce in March 1999.

2. Personify: This startup company offered reporting and analysis software that revealed patterns of visitor behavior, and consulting expertise to help companies act on those patterns to become more profitable. However, the company's Web literature was vague on the underlying technology for its products. Personify offered a hodge-podge of products, including click stream analyzers,

customer profile management, e-mail campaign management, and data warehouse/mart solutions. All these offerings pointed toward offline analysis, rather than real-time, sense-and-respond technology.

3. BroadVision: BroadVision offered e-commerce solutions (called "one-to-one" solutions) for various types of firms (B2B, B2C, etc.) to develop fully functioning e-commerce sites. BroadVision's products included content management, transaction processing, relationship management, billing, and business system integration with both front- and back-office systems. This company also sold a personalization component called "One-to-One Knowledge" that allowed retailers to create user profiles, observe and report on user choices, and define manual business rules on what content to deliver to a particular user.

Relationship Management Products

1. Siebel: At the time, Siebel was the largest provider of customer relationship management (CRM) products. This company offered several products related to overall customer management, including sales-force management, service-center management, call-center automation, and a marketing- and campaign-management tool. Siebel's marketing management tool, aimed at e-retailers, was heavily oriented toward online analytical processing (OLAP) and visualization. These tools were not geared for real-time recommendation.

2. E-Phiphany: This company sold a suite of products that centered around the company's approach of one-to-one marketing. The company's value proposition was to analyze Web data and then blend it with back- and front-office data for a complete view of customers and suppliers. E-Phiphany offered a datamining solution equipped with decision tree, clustering and scoring support, an OLAP tool, real-time, predictive personalization for Web sites and other customer-interaction channels. Most importantly, its solutions featured self-learning and self-adjusting customer targeting that discovered patterns in customer behavior and automatically adjusted to them in real-time.

3. Cogit: This startup company offered a value proposition that was very similar to SDC's. The company's products purportedly allowed retailers to link customers in both online and off-line worlds. Cogit's product (called RealProfile) created anonymous consumer profiles for the site by matching visitor online interests with off-line consumer information drawn from The Polk Company, a compiler of

consumer data for more than 100 years. Visitors were segmented according to more than 500 demographic and behavioral attributes, including family demographics, lifestyle admissions, products purchased, etc. Their products utilized *cookies* to reference visitors against an anonymous-profile database. Visitor profiles were then tracked using event tags on key on-site activities, such as responses to promotions or sweepstakes, purchases, home page visits, return visits, product categories or content channels, click-through from banner ads or e-mails. As part of its privacy protection initiatives, Cogit.com did not track consumer behavior across different client sites.

Other Products

1. Vignette: This company had its roots in Web content management and had extended its reach to the delivery of customized and personalized Web experiences through its acquisition of the analytic capabilities in DataSage. Vignette's flagship product was the Story Server (Vignette believed that each Web experience must be like a well-told story); customers could add a syndication component (for managing syndicated content), a "multi-channel" server to combine on- and off-line marketing, etc. While Vignette did not have a true personalization product, it entered into temporary original equipment manufacturers (OEM) agreements with Net Perceptions for this technology.
2. Oracle/Darwin: Darwin was Oracle's data warehousing product, built to integrate with a retailer's ERP, supply-chain, and front-office software for enterprise data mining. Oracle positioned Darwin as a full-featured datamining product, with extensive visualization and business-analysis tools. Depending on the type of datamining application, Darwin was equipped with algorithms to run decision trees, neural networks or memory-based reasoning.

Using this basic classification, the planning team created Figure 7.2 to display many of SDC's potential competitors.

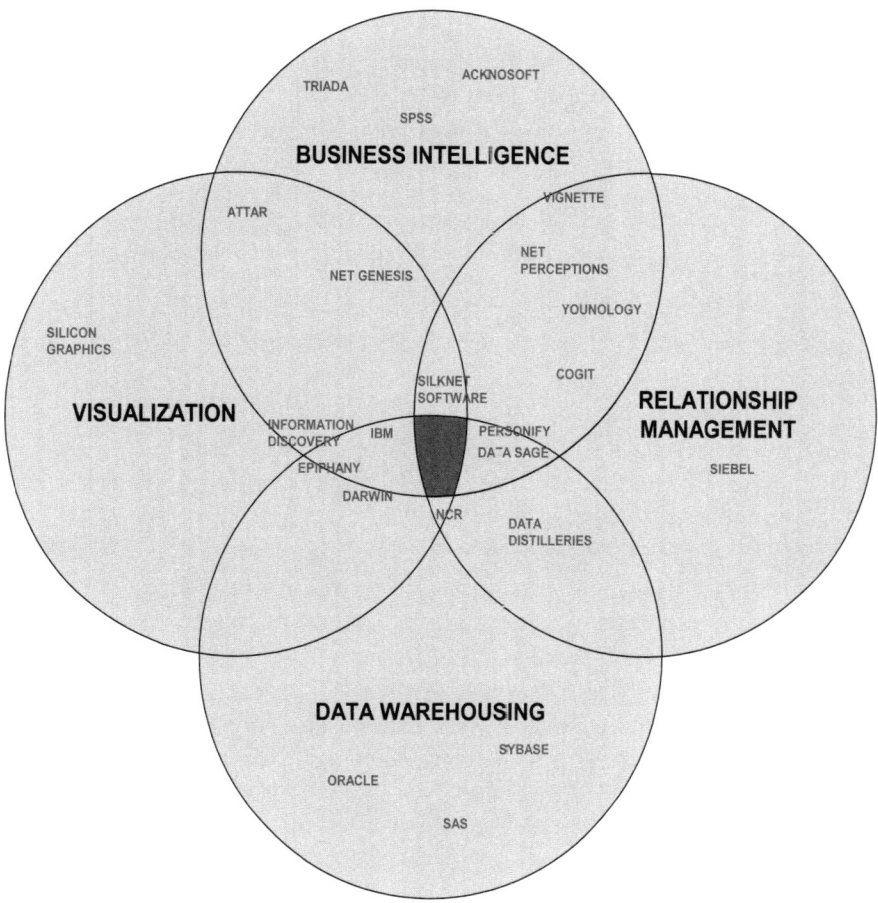

Figure 7.2. Potential Competitors and Features

7.2 Articulating the Critical Issues

The discussion of critical issues can begin anywhere. The critical-issues maps or grids are merely devices to help ensure a comprehensive elicitation of issues, and to keep track of what areas have been addressed.

7.2.1   Political Issues:

Patent Policy

Patent policy, in general, is an infrastructure issue where companies from many sectors and universities try to have their voice heard on the broad issues of encouraging innovation and protecting intellectual property. The e-commerce business ecosystem, in particular, had been affected greatly by the granting of general business-process patents such as Amazon's *One-Click* patent.

> United States Patent   5,960,411
>   September 28, 1999
> Method and system for placing a purchase order via a communications network
> Abstract
> A method and system for placing an order to purchase an item via the Internet. The order is placed by a purchaser at a client system and received by a server system. The server system receives purchaser information including identification of the purchaser, payment information, and shipment information from the client system. The server system then assigns a client identifier to the client system and associates the assigned client identifier with the received purchaser information. The server system sends to the client system the assigned client identifier and an HTML document identifying the item and including an order button. The client system receives and stores the assigned client identifier and receives and displays the HTML document. In response to the selection of the order button, the client system sends to the server system a request to purchase the identified item. The server system receives the request and combines the purchaser information associated with the client identifier of the client system to generate an order to purchase the item in accordance with the billing and shipment information whereby the purchaser effects the ordering of the product by selection of the order button.
> Inventors: Hartman, Peri (Seattle, WA); Bezos, Jeffrey P. (Seattle, WA); Kaphan, Shel (Seattle, WA); Spiegel, Joel (Seattle, WA)
> Assignee: Amazon.com, Inc. (Seattle, WA)
> Appl. No.: 928951
> Filed: September 12, 1997

That a patent could be issued for something so central and obvious chilled the e-commerce community. GNU organized boycotts to pressure Amazon.com into not trying to enforce its patent. Even though Amazon and Barnes and Noble settled their long-running lawsuit in March 2002, GNU continued the boycott pressure. Terms of the settlement were not revealed in a way that allowed GNU to assess victory over or defeat of the patent.

Our thoughts on patents had come a long way in just a few months from the early statement about patenting "remote-controlled, secure, residential, real-time datamining agents." We wanted to protect basic mechanisms for making recommendations from what we learned about both the individual and the relevant segment. In addition to general patents covering business processes, we encountered a slew of specific patents on recommendation engines.

> United States Patent   4,870,579   Hey
>   September 26, 1989
> System and method of predicting subjective reactions.
>
> United States Patent   4,996,642   Hey
>   February 26, 1991
> System and method for recommending items.
>
> United States Patent   5,704,017   Heckerman, et al.
>   December 30, 1997
> Collaborative filtering utilizing a belief network.
>
> United States Patent   5,749,081   Whiteis
>   May 5, 1998
> System and method for recommending items to a user.
>
> United States Patent   5,842,199   Miller, et al.
>   November 24, 1998
> System, method and article of manufacture for using receiver operating curves to evaluate predictive utility.
>
> United States Patent   5,867,799   Lang, et al.
>   February 2, 1999
> Information system and method for filtering a massive flow of information entities to meet user information classification needs.
>
> United States Patent   5,884,282   Robinson
>   March 16, 1999
> Automated collaborative filtering system.

United States Patent 5,918,014 Robinson June 29, 1999
Automated collaborative filtering in World Wide Web advertising.

United States Patent 5,983,214 Lang, et al. November 9, 1999
System and method employing individual user content-based data and user collaborative feedback data to evaluate the content of an information entity in a large information communication network.

United States Patent 6,014,654 Ariyoshi January 11, 2000
Information filtering apparatus for filtering information for interests of users and a method therefore.

United States Patent 6,018,738 Breese, et al. January 25, 2000
Methods and apparatus for matching entities and for predicting an attribute of an entity based on an attribute frequency value.

United States Patent 6,029,161 Lang, et al. February 22, 2000
Multi-level mindpool system especially adapted to provide collaborative filter data for a large-scale information filtering system.

United States Patent 6,041,311 Chislenko, et al. March 21, 2000
Method and apparatus for item recommendation using automated collaborative filtering.

United States Patent 6,049,777 Sheena, et al. April 11, 2000
Computer-implemented collaborative filtering based method for recommending an item to a user.

United States Patent 6,064,980 Jacobi, et al. May 16, 2000
System and methods for collaborative recommendations.

> United States Patent 6,092,049 Chislenko, et al.
> July 18, 2000
> Method and apparatus for efficiently recommending items using automated collaborative filtering and feature-guided automated collaborative filtering.
>
> United States Patent 6,108,493 Miller, et al.
> August 22, 2000
> System, method, and article of manufacture for utilizing implicit ratings in collaborative filters.
>
> United States Patent 6,112,186 Bergh, et al.
> August 29, 2000
> Distributed system for facilitating exchange of user information and opinion using automated collaborative filtering.
>
> United States Patent 6,266,649 Linden, et al.
> July 24, 2001
> Collaborative recommendations using item-to-item similarity mappings.

So the specific area in which our patents will be pursued is not characterized by seemingly overly broad patents such as One-Click, but rather by a complex pastiche of relatively narrow and seemingly overlapping patents. The good news is that with this many patents granted in the area, we should be able to carve out an approvable niche. The bad news is that any potential competitor could probably do the same. At the time of this planning exercise, it seemed clear that we could not be blocked from operating by other patents. Some final patents seemed likely to be approved, but, despite the uniqueness of our approach, nothing so broad as the One-Click Patent would come our way. Since then, the United States Patent and Trademark Office (USPTO) has granted our "Application To Make Special,"[83] reviewed and rejected our claims, which is pretty much the standard outcome if the patent attorneys have written the claims

---

[83]An "Application To Make Special" asserts that an expedited review could keep competitive claims from causing business problems in a sector. One tactic involves creating a subset of the claims into a separate patent application along with an "Application To Make Special." If the application is granted, the review of the subset of claims encourages the examiner, once familiar with the case, to review the entire set of claims at the same time. When justified, this leads to more efficiency in the examiner's effort, and shorter overall time to reach a conclusion.

broadly. SDC replied to the rejection within the six-month allowed frame. The USPTO rejected the reply, but accepted the priority date of the initial application. This means that our original filing date of November 1999 will be the starting date of patent protection, when SDC and the USPTO finally agree on the defensible claims.

The more broadly the independent claims are drafted, the more protracted this dance between the patent attorneys and the USPTO becomes. Patent attorneys specialize in being broad and vague in their writing. This is perhaps why one of our patent attorneys once said, in a striking moment of candor, that you shouldn't hire patent attorneys for any other legal work.

Venture-capital firms seem to place great emphasis on patents in their funding decision. In many areas I'm sure it is justified. With respect to technology-enabled marketing, I'm not as convinced. As indicated above, the pastiche of patents in this area disposes me to think we will find a patent compromise that will protect our efforts, but will not be a barrier to competitive entry.

Privacy

The public-policy debate concerning privacy intensified during this time. While U.S. policy makers seemed less strident in their proclamations than their European counterparts,[84] either legislation or the bully pulpit seemed likely to keep attention focused in this area. We felt protected by this attention. Our approach to personalization required far less invasion of protected areas than any other approach. A few major competitors were blocked from pursuing some of their more aggressive strategies because of the debate. Engage.com had 800 pieces of information gathered on something like 55 million Internet users, but was constantly on the defensive, claiming all profiles were anonymous. Ever since Doubleclick purchased Abacus Direct with its 90 million-person database, privacy hawks predicted Doubleclick would combine its cross-Web site tracking capability with the extensive information in Abacus to become a direct-marketing leviathan. When Doubleclick whispered that intention, the press outcry was so immediate and

---

[84]Swire, Peter P. and Robert E. Litan, *None of Your Business: World Data Flows, Electronic Commerce, and the European Privacy Directive*, Boston: The Brookings Institute (1998).

intense it backpedaled and hired a chief-privacy officer to calm things down.

Our privacy-friendly approach was an asset. We fell under the traditional marketing maxim of *know your customers*. E-commerce sites and virtual malls were to be our clients. Their customers opted in, and with only ZIP code and gender our learning algorithms could make offers more likely to result in sales. Profit to our clients meant revenue for us.

Emergence of Secure Payment Systems

Trust in the security of Internet transactions is an area in which attitudes have changed rapidly. When I started teaching about the digital economy in 1996, very few people were willing to give out credit-card information over the Internet. Remember that Amazon's predecessor to the *One-Click* patent involved methods for making secure transactions over insecure networks. The patent, by the way, involved little more than typing in some information and completing the transaction by phone. While people routinely allowed unknown waiters to walk off with their credit cards, the idea of typing that same number into a computer frightened most people. Secure-socket layer (SSL) and HTTPS domains were in their infancy. The likely diffusion of this technology would have a positive impact on the e-commerce market in general, rather than a specific boost to the personalization or knowledge-management sector in particular.

Taxation of E-Commerce

More than 30,000 local taxing authorities look at potential tax revenue from e-commerce and wonder, "Why not?" After all, bricks-and-mortar stores pay local sales taxes, so why should e-retailers be exempt? But e-commerce is extremely mobile, lucrative, and efficient while not generating any negative externalities such as pollution. It would also be very difficult to determine how much e-retailers should be taxed. Problems exist with the ability of governments to track and collect taxes in a digital economy. Concealed trade is made possible by cryptographic tools. In addition, governments would have difficulty establishing tax jurisdiction because physical location of transacting parties is no longer clearly defined in electronic commerce. Source and residence definitions could be constituted as the location of a transaction server or of a Web site, as opposed to that of people or businesses. A transaction could pass through multiple jurisdictions. Congress has so far resisted the temptation to tap this nascent sector for needed revenue, but the pressure remains. While the near-term likelihood of e-commerce tax is small, the impact if enacted would we widely felt on the whole sector, rather that on the personalization or knowledge-management areas specifically.

7.2.2 Behavioral Issues

With consumer products, the company-level behavioral issues usually involve the traditional product-development research cycle for new products or customer-satisfaction research for existing products. While PersonalClerk ultimately served consumers, it was a business-to-business product. SDC's direct clients were the e-commerce sites. Thus, company-focused behavioral issues dealt more with how our hardware/software platforms were integrated into the e-commerce infrastructure of our clients.

Many e-commerce sites have a general fear of technical integration with third parties. While a whole-product solution often requires collaborative efforts, and best-of-breed solutions often come from separate, specialized companies, a history of trying to support legacy code from failed software companies makes clients wary of engaging with startups. That's one reason that Bizrate.com wanted a code escrow. It's one possible justification for Yahoo's! well-known "not-invented-here" syndrome. To get in the game, we needed a system that could not possibly crash the client's site. That had to be obvious from the design, but was not hard to achieve. Beyond allaying the fear of disaster, we had to provide the technical support in installation and client-service support in use and maintenance that helped the client gain the desired benefits from our system.

Our ability to deal with the need for tech support and client-service support came from our already assembled team and our access to the UCLA talent pool. Our human capital represented an outstanding resource for SDC.

The behavioral issues affecting the business ecosystem concern e-shoppers' expectations of the online experience. This goes beyond the issues of trust in e-commerce transactions. As pointed out previously, in a report on *Knowledge Management*, IDC believes personalization/customization will be the "ante" for successful e-commerce sites. "Why? Because it works." Customers learn to expect personalized experiences from the e-commerce sites they patronize.

The infrastructure-level behavioral issues center on the growth in the online population, the growth in the time and money Netizens spend online. In the summer of 2000, approximately 53 million U.S. households (51%) were online, spending an average of 975 minutes online per month.[85] The U.S. Department of Commerce estimates that $5.5 billion in sales occurred online in the second quarter of 2000, with 15% quarter-to-quarter growth. While growth was volatile and widely heralded failures occurred in 2000, online retail sales did grow 21% (annual percentage rate) to $51.3 billion for all of 2001, and are expected to jump 41% to $72.1 billion in 2002, based on the results through the first quarter of 2002.[86] To repeat what was said in Chapter 4, email marketing grew to $927 million in 2001, up 87% from 2000. Cost of a sale to an existing customer through email was $1, compared to $20 for equivalent customers through direct mail.[87]

The behavioral fundamentals for the business ecosystem and the infrastructure of e-commerce were sound. We could expect the growth in the online audience to increase the size of the general e-commerce market and consumers' expectations of a more tailored and personalized experience to increase the share of the e-commerce pot that was served by personalization engines.

7.2.3   Economic Issues

At the company level, the revenue-side economic issues dealt primarily with SDC's access to venture capital and to major clients. Fisher Pennington (FP) were key to both of these issues. Both Internet advertising and e-commerce were concentrating into the hands of fewer and fewer major players. FP's connections with MSN, AOL, and Yahoo! were obviously going to be critical to SDC's success. While our dependence on FP grew, that firm had come through for SDC in major ways. I saw little alternative to Bud's

---

[85]The fact in this section are from The Jupiter Consumer Survey, Volume 4 "US Online Demographics: Fundamentals and Forecasts," Spring 2000.
[86]*The State of Retailing Online 5.0: Performance Benchmark Report.* June 2002 Shop.org
[87]Schoenberger, Chana R. (2002), "Marketing: Web? What Web?" Forbes Online, June 10, 2002.

growing importance. We were part of a *Kiretsu* in which Bud in particular, and FP in general, were the shapers of the economic web.[88]

Our focus was on getting quickly to market with some early wins that were referenceable accounts. The venture funding was what allowed us to handle the costs associated with building the technology team and client-services staffs needed to handle the anticipated demand. Our revenue forecast was based on relatively balanced contributions from e-commerce sites and ad-revenue based sites, with revenue from email optimization building more slowly. I was concerned that ad revenue was highly concentrated in the top sites, while ad opportunities were almost unlimited. As seen in the table below, the top sites were still doing very well in June 2000, but these four sites represented perhaps 50% of all Internet ad revenues that month. The potential for an imbalance between the publishers' supply of Internet ad space and the advertisers' demand for something more than *eyeballs* (e.g., purchase, subscription, or measurable branding effects) left unanswered questions about the viability of the CPM model that dominated ad revenue compared to the CPC model. Rate cards for these top sites ran from $42 to $22 CPM, while sites just below the top 10 had difficulty attracting advertising at any rate. Our revenue projects from Internet advertising were based on bumping up effectives of banner ads (and other ad formats), thus justifying higher rates, but the imbalance between supply and demand threatened the ability of publishers to sustain even the rates they were currently charging. E-commerce malls such as those of MSN, AOL, and Yahoo! were important targets for SDC, but the plan was to approach them after deployment in easier client environments had been successful.

Company	June 2000 Ad Revenues
Yahoo	$ 107,129,800
Lycos	$ 73,574,000
Microsoft	$ 69,242,000
AOL	$ 68,233,100

---

[88]For an early discussion of shaping economic webs see Hagel, John III, "Spider versus spider," *The McKinsey Quarterly*, 1996, number 1.

7.2.4 Sociological Issues

The company-level sociological issues concerned building a culture of mutual support, trust, and volunteerism. We sought to fill the ranks of both tech and management staffs primarily through UCLA alumni and secondarily through the network of current employees. This helped ensure both A-level players and employees who could integrate quickly into the emerging culture. It helped insulate us, to a limited extent, from the high-flying tech superstars of the broader business ecosystem. We paid market rates for top talent, but wanted employees who fit into the vision of the company, rather than large egos who wanted the company to fit around their needs.

The sociological issues in the infrastructure almost always concern the effects of the baby boom and its echoes. Exactly where the baby boomers were in their earning cycle and family cycle had an impact on the growth of the online population. Two-wage-earner families with more available money than available time were pushing growth in the online population as a convenience in their crowded lives. PersonalClerk fit right into this trend.

7.2.5 Technological Issues

The business ecosystem provided technical standards that greatly aided our efforts: Structured Query Language (SQL) for our database technology, Linux operating systems as an inexpensive OS platform, and Intel Architecture as a foundation for the pizza-box hardware. Our tech-team members were masters of these components.

To this complete technology pallet I added marketing-science models that were easily implemented. ZipSegments resulted from my reanalysis of the ZIP code files from the 1990 U.S. Census. It was done and available, and would last at least another two years, until the 2000 Census files would be available. Segment-based learning algorithms were straightforward complements to ZipSegments. I had an experienced sense of which marketing models and methods were practical in our circumstance. ZipSegments, marketing-science models, and the intellectual capital in the advisory panel gave SDC the flexibility of a multi-channel marketing scheme. If the e-commerce side of the venture ran into tougher times, that flexibility could open efforts in other technology-enabled marketing arenas.

7.2.6 The Key Decision

The central strategic decision in the effort concerned whether SDC should continue its broad emphasis on technology-enabled marketing optimization (TEMO), as characterized by PersonalClerk, or narrow its focus to Internet advertising, as pushed by the new CEO. The time frame was the fall of 2000, and the implications were evaluated both in terms of the likelihood of success of SDC and the resulting valuation of the company.

7.3 Mapping the Critical Issues

From the list of issues in the previous section, a mental map of the strategic terrain may be constructed. In this effort, the listed issues are connected in a heuristic what-influences-what sense. The overall map appears in Figure 7.3. While I built the mental map a section at a time, essentially as I wrote Section 7.2, telling the story is easier from the most direct influences to the more remote.

Figure 7.3. The Mental Map of Factors Affecting SDC's Success

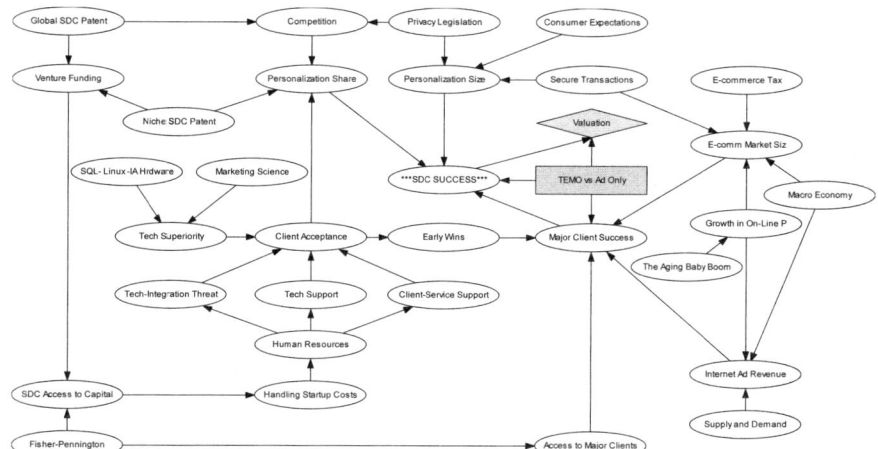

Consider the arrows that point directly to the node "SDC Success." The success of SDC seemed most directly affected by both the size of the personalization market and SDC's share of that market. A big share of a big personalization market leads to a high likelihood of a big win. A small share of a big market or a big share of a small market would more likely lead to a steady state for the company, whereas a small share of a small market would most likely lead to death. Success with major clients would have a direct impact on the success of SDC. These oval nodes are *chance nodes* characterized by answers to general questions such as, "What are the chances that success with major clients leads to overall SDC success?" But the meaning of success depends on the strategic decision between the broader TEMO focus and the narrower emphasis on Internet ads. Before the crash of the Internet advertising market, the SDC board generally believed that a big win in the advertising arena had more up side than a big win in the broader e-commerce market. I reflected this in the Valuation (the diamond-shaped *utility node*) discussed in Section 7.4 below. The rectangular *decision node* (TEMO vs. Ad Only) points at Major Client Success, SDC Success, and Valuation, reflecting the direct impact of this strategic decision on these nodes.

The multi-channel-marketing capability, built into the original design of ZipSegments, gave SDC opportunities for success even if the e-commerce sector suffered decline. SDC's share of the personalization market would be affected by the strength of the competition and the likelihood of SDC obtaining niche patent protection. Global patent protection for SDC would weaken the competition, as would an increased emphasis on privacy legislation. Privacy legislation, along with progress on secure transactions and consumers' expectations for personalization, would affect the size of the personalization market. On the TEMO side, major client success would most likely be affected by access to those major clients, early wins with other referenceable accounts, and the overall size of the e-commerce market. The overall size of the e-commerce market is affected by progress on transaction security, the likelihood of holding off the sales tax on e-commerce transactions, the growth in the online population, and the state of the macro economy. Large-scale demographic shifts, such as the aging of the baby boom, would affect the growth in the online population. Early wins would be the direct result of client acceptance, which also would affect SDC's share of the personalization market. Client acceptance would be directly affected by SDC's technical superiority, tech support, and client-service support, as well as by the perceived threat from technical integration. In different ways, SDC's intellectual capital would drive these factors. Our access to human resources in computer science and management would drive our ability to handle the technical integration threat, staff tech support and client services. It also would drive SDC's technical superiority through mastery of the standard SQL databases, Linux OS and Intel hardware (as long as those standards were stable), and SDC's foundation in marketing science. But the A-list personnel needed for this mastery depended on our ability to handle the start-up costs. Fisher Pennington had a major influence over our access to key clients and capital. Obviously, our desirability to venture funds would affect our access to capital. The venture-fund outlook would be affected by the likelihood of either global or niche patents.

On the Internet ad-market side, the influences discussed above are still at play. The personalization approach works here, since Internet ads are better received if they are anticipated, personal, and relevant.[89] The Internet ad market, however, face the issue of the supply-demand balance, and the impact of that balance (or lack thereof) on ad revenue. The ability to add advertising is practically unlimited. We

---

[89]Godin, Seth, and Don Peppers (1999), *Permission Marketing: Turning Strangers Into Friends, and Friends into Customers*, New York: Simon & Schuster.

know that in the very long term, advertising spending is a mean-reverting process that represents about 2.15% of GDP.[90] That long-term value tolerates wide swings. We also postulate that there is a long-term pressure to shift ad spending to being proportional to the amount of leisure time consumers spend on the Internet. These factors affect the ability of clients to monetize the increase in advertising effectiveness achieved through SDC technology, thus influencing the likelihood of major client success in this space.

7.4 Valuation

The Valuation (the diamond-shaped *utility node*) has three states: a *big win* reflecting the full weight of the speculated corporate valuations for the next round of funding, *steady state* to indicate a cash-flow neutral position that didn't require addition capitalization, and *death*. *Death* of the firm was valued at a loss of that capital invested at that point (–$6.25 million). This value was obviously the same whether the broader TEMO emphasis or the narrower Ad Only emphasis was selected. The *steady state* for either side of the strategic decision was valued at the amount of capital put into the enterprise ($6.25 million). The value of a *big win* was where differences existed. The second rumored valuation from Goldman Sachs in the fall of 2000 was between $50 and $100 million. This range corresponded to independent discussions I held with a very major investor in the B-Round. In a nod to the prevailing feeling that a *big win* on the Ad-Only side had more up-side potential than the broader TEMO emphasis, I placed that value at the upper end, and the *big win* with TEMO at the lower end of this valuation.

7.5 Plans Must Be Dynamic

Things change. Elsewhere, to motivate this strategic-marketing-planning approach, I said, "If you make *change* your enemy, you will always be at war."[91] The continuous stream of events renders a written plan obsolete as soon as it comes off the printer. Some

---

[90] I've heard numerous people speak of mean reversion and this long-term average as well known, but the only explicit reference I've found is: Kornelis, Marcel (2002), "Modeling Advertising Markets Using Time-Series Data" doctoral dissertation, Rikjsuniversiteit, Groningen, The Netherlands, p.115.
[91] Cooper, Lee, Troy Noble, and Elizabeth Korb (1999), "Strategic Marketing Planning in Turbulent Environments: the Case of PromoCast," *Canadian Journal of Marketing Research*, 18, 46-66.

management gurus use this as a justification for failing to plan – reminding us of the old saw that those who fail to plan plan to fail. I use the immediate obsolescence of written plans to justify dynamic planning. We have already taken the first step in making plans dynamic by talking about the relationship between events in terms of their likelihood. Whatever the likelihood of events in the plan, once the event occurs its likelihood changes to 1.0 if it happens as you anticipated or 0.0 if you guessed wrong. If you are telling stories about possible futures, as done in scenario planning,[92] you have a whole sequence of events that lead to a scenario outcome. If one event in the sequence unfolds in an unanticipated manner, you are left to guess what impact that has on the overall story.

The mental map I presented in the previous section is really a network of beliefs I hold about the interrelations of market forces and the consequences of strategic decisions. If I add a little discipline about how I state my beliefs, I can create a dynamic belief structure, called a Bayesian belief network, that allows me to update information when the likelihood changes and assess the impact of any proposed change on the likely success of my venture.[93] The structure of a Bayesian Belief Network requires you only to ask questions about the local relations (i.e., the arrows pointing directly to a particular node are the only ones whose likelihood you must assess). The rest is mathematics, which others have developed elsewhere.[94] The nodes with no arrows pointing to them are *parent nodes*, such as the likelihood of SDC obtaining a niche patent. For these *parent nodes*, you must specify the states (niche protection granted or denied, in this example) and your initial guess at the likelihood of these outcomes (95% chance of obtaining a niche patent and 5% chance of being denied). Having just the reverse of these likelihoods as my guess at SDC's chances for a global patent reflects in numbers what I discussed at the beginning of Section 7.2.1. These are rough guesses, but that is all that is needed in the

---

[92]Schwartz, Peter (1996) *The Art of the Long View*, New York: Currency Doubleday.
[93]Bayesian Belief Networks are *directed acyclic graphs* (DAGs). They are *directed* in the sense that the arrows indicating causal direction all flow one way. They are *acyclic* in the sense that the arrows do not circle back on themselves. This eliminates what are commonly called feedback loops. But since the notion of current events affecting the past is the realm of fiction rather than science, we are not actually inconvenienced by this restriction. We can represent repeated processes over time with *feed-forward* influences.
[94]Pearl, Judea (2000), *Causality: Models, Reasoning, and Inference*, Cambridge, UK: Cambridge University Press.

beginning. The states and likelihoods I assigned to all the parent nodes in Figure 7.3 are listed in Table 7.1. For each of these nodes I listed only two states. There is no limit on the number of states, but it's best to keep it simple to start with, since the number of columns in a table is the product of the number of states in each node pointing into the node for that table. This table reflects what I believe: consumers have grown to expect a more personal experience in their interaction with their often-visited sites on the Internet (0.70); Congress will continue its near-term waiver of taxes on the Internet transactions (0.95); I will continue to be able to connect SDC with the marketing-science expertise it needs to move forward (0.98); privacy legislation will continue to be a hot issue (0.70); stable standards will continue for database software (SQL), operating systems (Linux), and Web servers (Intel Architecture) (0.90); the standards for secure transactions will continue to feed customers' perceptions of high security (0.8); and Fisher Pennington will continue to focus attention on the needs of SDC (0.80).

Table 7.1. Likelihood of States in Parent Nodes.

Consumer Expectations	Initial Likelihood
Expect Personalization	80%
Low Expectations	20%
E-Commerce Tax	Initial Likelihood
Near-Term Waiver	95%
Open to Tax	5%
Global SDC Patent	Initial Likelihood
Denied	95%
Global Protection	5%
Marketing Science	Initial Likelihood
Available	98%
Unavailable	2%
Niche SDC Patent	Initial Likelihood
Denied	5%
Niche Protection	95%
Privacy Legislation	Initial Likelihood
Backburner	30%
Hot Issue	70%
SQL-Linux-IA Hardware	Initial Likelihood
Stable Standards	90%
Uncertain Standards	10%
Secure Transactions	Initial Likelihood
High Perceived Security	80%
Low Perceived Security	20%

The Aging Baby Boomers	Initial Likelihood
Pushes Growth	95%
Inhibits Growth	5%
FP	Initial Likelihood
Continued Attention	80%
Waning Interest	20%

The simplest joint table is for a node with one arrow pointing to it. In this mental map in Figure 7.3, for example, only one factor affects SDC's ability to handle the perceived *threat from technical integration* (B1). If the *human resources* are available, then I judge the likelihood that the client would perceive a high technical-integration threat to be 20%, with a complementary 80% chance that the client would perceive a low threat. If the human resources are not available to handle this, I estimate the likelihood of perceiving a threat from the technical integration climbing to 70%. All of the two-way conditional tables appear in Table 7.2. Human resources also drive SDC's ability to provide tech support (B2). If the human resources are available, I rated the likelihood of tech support being rated as good at 0.90. If the human resources are unavailable, the tech support rating rests much more heavily on the reliability and availability of the underlying technology, and the likelihood of having tech support rated as good drops to 0.55. A more extreme version of this reasoning applies to the impact of human resources on client-service support (B3). If the human resources are available, the likelihood of a good rating is 0.9. If unavailable, the likelihood drops to 0.40.

Table 7.2. Two-Way Conditional Likelihoods.

Tech-Integration Threat (B1)		Conditional Likelihood	
Human Resources		Available	Unavailable
	High	0.2	0.7
	Low	0.8	0.3
Tech Support (B2)		Conditional Likelihood	
Human Resources		Available	Unavailable
	Good	0.9	0.55
	Poor	0.1	0.45
Client-Service Support (B3)		Conditional Likelihood	
Human Resources		Available	Unavailable
	Good	0.9	0.4

		Poor	0.1	0.6
Handling Startup Costs (E6)			Conditional Likelihood	
SDC Access to Capital			Likely	Unlikely
		Likely	0.99	0.01
		Unlikely	0.01	0.99
Human Resources (B5)			Conditional Likelihood	
Handling Startup Costs			Likely	Unlikely
		Available	0.9	0.01
		Unavailable	0.1	0.99
Access to Major Clients (E3)			Conditional Likelihood	
FP			Continued Attention	Waning Interest
		Likely	0.95	0.7
		Unlikely	0.05	0.3
Growth in Online Pop (B6)			Conditional Likelihood	
The Aging Baby Boom			Pushes Growth	Inhibits Growth
		Sustained Growth	0.7	0.4
		Flat	0.3	0.6
Early Wins (E5)			Conditional Likelihood	
Client Acceptance			Likely	Unlikely
		Likely	0.9	0.2
		Unlikely	0.1	0.8

The judgments are no more difficult, but the language is slightly more convoluted when the states are labeled *likely* and *unlikely*, as in the fourth table (E6) in Table 7.2 where "handling startup costs" is described as *likely* or *unlikely*. So the linguistic twist is that you need to assess how *likely* it is that "handling startup costs" is properly described as *likely*.

By the time you get to the three-way conditional tables (nodes that have three arrows pointing to them) you should realize the value of keeping things simple. The number of columns in these tables will equal the product of the number of states in each of the nodes pointing into this node. The number of rows corresponds to the number of states in this particular node. The first section of Table 7.3 deals with the impact of patents on the prospect for venture funding (E3). A general tactic for filling in such tables with first guesses is to first find the most positive column. In this case, the most positive condition would be if patents provided both *global protection* and *niche protection*. The question to answer here is, "If patents provided both global protection and niche protection, what are the odds that venture funding is likely?" While I don't expect a global patent to be issued, it would put SDC in an extremely favorable position. I put the odds of venture funding in this best case at 99 in 100 (0.99 to 0.01). In the earlier discussion I indicated I believed that either niche protection or global protection by patents would enhance the likelihood of venture funding. If neither patent umbrella were forthcoming, the prospects for venture funding become dim. I believe a good first guess is that the odds for venture funding being *likely* in this worst case are one in 20 (0.05 to 0.95). From the best case for SDC to the worst case for SDC, we could order the columns. In the event that global patent protection was offered but specifics were denied, SDC would be more likely to get venture funding than if global patents were denied and niche patents were granted. Rough, ordinal bounds on the judged likelihoods are not difficult to establish by such ordering. While the final likelihoods will be more accurate when the likelihoods of these conditions are better assessed, even the rough guesses are useful, as we will see. Consider the second section in Table 7.3. This table (P4) concerns the likely impact of a global patent for SDC and the prospects for privacy legislation on the strength of SDC's competition in the personalization space. Again we find the most positive column. In this case, the best prospect for SDC is if global patent protection is granted to SDC and privacy legislation is a hot issue. If these conditions hold, what is the likelihood that the competition for SDC is *weak*? I judged that likelihood to be 0.9, with a complementary 0.1 likelihood that the competition would be *strong*.[95]

---

[95] Formally, such judgments are called *ratio-scale* judgments, since an event with a judged likelihood of 0.5, for example, should be twice as likely as one with a judged likelihood of 0.25. The sum of the likelihoods assigned in a particular column must be 1.0, and the ratios between rows in that column should reflect the real ratios of the likelihood of these states. To begin with, a lot of error is likely to exist in these judgments. But, as I will continue to emphasize, we need only a place to begin and a way to get better. As experience grows, and/or specific research projects are undertaken to assess the needed likelihoods more accurately, we have a way to be better.

Table 7.3. Three-Way Conditional Likelihoods.

Venture Funding (P3)		Conditional Likelihood		
Global SDC Patent		Denied		Global Protection
Niche SDC Patent	Denied	Niche Protection	Denied	Niche Protection
Likely	5%	70%	95%	99%
Unlikely	95%	30%	5%	1%
Competition (P4)		Conditional Likelihood		
Global SDC Patent		Denied		Global Protection
Privacy Legislation	Back Burner	Hot Issue	Back Burner	Hot Issue
Weak	70%	60%	80%	90%
Strong	30%	40%	20%	10%
SDC Access to Capital (E4)		Conditional Likelihood		
Venture Funding		Likely		Unlikely
FP	Continued Attention	Waning Interest	Continued Attention	Waning Interest
Likely	95%	80%	90%	1%
Unlikely	5%	20%	10%	99%
Tech Superiority (T3)		Conditional Likelihood		
Marketing Science		Available		Unavailable
SQL-Linux-1A Hrdw	Stable Standards	Uncertain Standards	Stable Standards	Uncertain Standards
Obvious Superiority	90%	85%	70%	40%
Confused Messages	10%	15%	30%	60%

The rest of the conditional probabilities appear in Tables 7.4 – 7.6. Filling these out can be a tedious exercise. Each column, however, asks a simple conditional question, "What are the likelihoods of these specific outcomes given a set of specific conditions?" In scenario planning you are left with a single question, "What is the likelihood of this story?" To be fair, scenario planning is more about envisioning possibilities than assessing likelihoods. But this begs the question of what we do about the possibilities once envisioned. So we are swapping a few unanswerable questions for numerous answerable ones, and we gain the ability to do something much more with the answer – assess the likelihood of the base scenario, find the nodes that most influence the final outcomes, run any what-if scenario, update the base scenario as speculation becomes reality or as research more precisely specifies the likelihood of events, and assess the expected value of any strategic decisions to be made.

Table 7.4. Four-Way Conditional Likelihoods.

Internet Ad Revenue (E9)		Conditional Likelihood			
Macro Economy		Recession			
Growth in Online Population		Sustained growth		Flat	
Supply and Demand		S_D Balance	Over supply	S_D Balance	Over supply
Growing		80%	40%	30%	1%
Shrinking		20%	60%	70%	99%
		Conditional Likelihood			
Macro Economy		Continued Boom			
Growth in Online Population		Sustained Growth		Flat	
Supply and Demand		S_D Balance	Over supply	S_D Balance	Over supply
Growing		75%	45%	60%	30%
Shrinking		25%	55%	40%	70%

Table 7.5. Five-Way Conditional Likelihoods.

E-Commerce Market Size (P9)		Conditional Likelihood			
Macro Economy		Recession			
Growth in Online Pop.		Sustained Growth			
E-Commerce Tax		Near-Term Waiver		Open to Tax	
Secure Transactions		High Perceived	Low Perceived	High Perceived	Low Perceived
Growing		70%	65%	60%	55%
Static		30%	35%	40%	45%
		Conditional Likelihood			
Macro Economy		Recession			
Growth in Online Pop.		Sustained Growth			
E-Commerce Tax		Near-Term Waiver		Open to Tax	
Secure Transactions		High Perceived	Low Perceived	High Perceived	Low Perceived
Growing		30%	25%	20%	15%
Static		70%	75%	80%	85%

Macro Economy		Conditional Likelihood		
Growth in Online Pop.		Continued Boom		
E-Commerce Tax		Sustained Growth		
	Near-Term Waiver		Open to Tax	
Secure Transactions	High Perceived	Low Perceived	High Perceived	Low Perceived
Growing	90%	85%	80%	75%
Static	10%	15%	20%	25%

Macro Economy		Conditional Likelihood		
Growth in Online Pop.		Continued Boom		
E-Commerce Tax		Flat		
	Near-Term Waiver		Open to Tax	
Secure Transactions	High Perceived	Low Perceived	High Perceived	Low Perceived
Growing	67%	55%	40%	35%
Static	33%	45%	60%	65%

Client Acceptance (B4)		Conditional Likelihood		
Tech Superiority		Obvious Superiority		
Client-Service Support		Good		
Tech Support	Good		Poor	
Tech-Integration Threat	High	Low	High	Low
Likely	90%	98%	85%	90%
Unlikely	10%	2%	15%	10%

		Conditional Likelihood		
Tech Superiority		Obvious Superiority		
Client-Service Support		Poor		
Tech Support	Good		Poor	
Tech-Integration Threat	High	Low	High	Low
Likely	85%	95%	60%	70%
Unlikely	15%	5%	40%	30%

		Conditional Likelihood		
Tech Superiority		Confused Messages		
Client-Service		Good		

Support				
Tech Support	Good		Poor	
Tech-Integration Threat	High	Low	High	Low
Likely	60%	65%	45%	55%
Unlikely	40%	35%	55%	45%

Conditional Likelihood
Tech Superiority
Client-Service Support

Confused Messages
Poor

Tech Support	Good		Poor	
Tech-Integration Threat	High	Low	High	Low
Likely	45%	50%	30%	40%
Unlikely	55%	50%	70%	60%

SDC SUCCESS (C1)

Conditional Likelihood

TEMO vs. Ad Only
Personalization Size — Larger

	TEMO			
Major Client Success	Likely		Unlikely	
Personalization Share	Low	High	Low	High
Death	0.1	0.05	0.5	0.2
Steady State	0.2	0.15	0.4	0.5
Big Win	0.7	0.8	0.1	0.3

Conditional Likelihood

TEMO vs. Ad Only
Personalization Size — Smaller

	TEMO			
Major Client Success	Likely		Unlikely	
Personalization Share	Low	High	Low	High
Death	0.2	0.1	0.7	0.6
Steady State	0.4	0.2	0.25	0.3
Big Win	0.4	0.7	0.05	0.1

Conditional Likelihood

TEMO vs. Ad Only
Personalization Size — Ad Only
Larger

Major Client Success		Likely		Unlikely	
Personalization Share	Low	High	Low	High
Death	0.1	0.1	0.8	0.8
Steady State	0.2	0.2	0.15	0.15
Big Win	0.7	0.7	0.05	0.05

	Conditional Likelihood				
TEMO vs. Ad Only		Ad Only			
Personalization Size		Smaller			
Major Client Success		Likely		Unlikely	
Personalization Share	Low	High	Low	High
Death	0.2	0.2	0.9	0.9
Steady State	0.4	0.4	0.09	0.09
Big Win	0.4	0.4	0.01	0.01

Table 7.6. Six-Way Conditional Likelihoods.

Major Client Success (E8)		Conditional Likelihood			
TEMO vs. Ad Only		TEMO			
Internet Ad Revenue		Growing			
E-Commerce Market Size		Growing			
Access to Major Clients		Likely		Unlikely	
Early Wins		Likely	Unlikely	Likely	Unlikely
	Likely	90%	60%	40%	10%
	Unlikely	10%	40%	60%	90%
	Conditional Likelihood				
TEMO vs. Ad Only		TEMO			
Internet Ad Revenue		Growing			
E-Commerce Market Size		Static			
Access to Major Clients		Likely		Unlikely	
Early Wins		Likely	Unlikely	Likely	Unlikely
	Likely	70%	40%	20%	5%
	Unlikely	30%	40%	80%	95%
	Conditional Likelihood				
TEMO vs. Ad Only		TEMO			
Internet Ad Revenue		Shrinking			
E-Commerce Market Size		Growing			

Access to Major Clients		Likely		Unlikely	
Early Wins		Likely	Unlikely	Likely	Unlikely
	Likely	80%	55%	35%	5%
	Unlikely	20%	45%	65%	95%
		\multicolumn{4}{c}{Conditional Likelihood}			

TEMO vs. Ad Only	TEMO
Internet Ad Revenue	Shrinking
E-Commerce Market Size	Static

Access to Major Clients		Likely		Unlikely	
Early Wins		Likely	Unlikely	Likely	Unlikely
	Likely	60%	35%	15%	2%
	Unlikely	40%	65%	85%	98%
		\multicolumn{4}{c}{Conditional Likelihood}			

TEMO vs. Ad Only	Ad Only
Internet Ad Revenue	Growing
E-Commerce Market Size	Growing

Access to Major Clients		Likely		Unlikely	
Early Wins		Likely	Unlikely	Likely	Unlikely
	Likely	90%	40%	20%	5%
	Unlikely	10%	60%	80%	95%
		\multicolumn{4}{c}{Conditional Likelihood}			

TEMO vs. Ad Only	Ad Only
Internet Ad Revenue	Growing
E-Commerce Market Size	Static

Access to Major Clients		Likely		Unlikely	
Early Wins		Likely	Unlikely	Likely	Unlikely
	Likely	90%	40%	20%	5%
	Unlikely	10%	60%	80%	95%
		\multicolumn{4}{c}{Conditional Likelihood}			

TEMO vs. Ad Only	Ad Only
Internet Ad Revenue	Shrinking
E-Commerce Market Size	Growing

Access to Major Clients		Likely		Unlikely	
Early Wins		Likely	Unlikely	Likely	Unlikely
	Likely	30%	5%	5%	1%
	Unlikely	70%	95%	95%	99%
		\multicolumn{4}{c}{Conditional Likelihood}			

TEMO vs. Ad Only	Ad Only
Internet Ad Revenue	Shrinking
E-Commerce Market Size	Static

| Access to Major Clients | | Likely | | Unlikely | |
Early Wins		Likely	Unlikely	Likely	Unlikely
	Likely	25%	2%	2%	1%
	Unlikely	75%	98%	98%	99%

7.6 What If?

We are now prepared to simulate the expected consequence of any scenario, and find the expected value of the strategic options. In this sense, the Bayesian Belief Network in Figure 7.2 is akin to a giant decision tree. *Giant* is a substantial understatement, since this network can take on more than 6 billion states.

The first *what if* is, "What if everything goes exactly to plan?" If everything follows the baseline probabilities depicted in the conditional tables, the aggregate chances for *death*, *steady state*, and *big win* are 0.35, 0.23, and 0.43, respectively. This reflects the still-rosy outlook of the fall 2000. The optimal decision is essentially a tie, with either strategic option producing an expected value for SDC of approximately $28 million, $3 million beyond the post-money valuation of the B-Round in the spring of 2000. Selecting the TEMO approach yields chances for *death*, *steady state*, and *big win* of 0.18, 0.26, and 0.55, respectively, while selecting the Ad-Only approach yields much higher chances for death (0.51, 0.19, and 0.30 for *death*, *steady state*, and *big win* respectively). So the Ad-Only approach is higher risk, but the higher potential return offsets the risk under the baseline scenario.

Investigating the nature of the higher risk is very revealing. If the growth in the e-commerce market flattens out, TEMO narrowly becomes the best choice (25 to 24) due to the multi-channel capability of the TEMO approach. TEMO is the best choice if the Supply-Demand balance in Internet ads tips to oversupply (28 to 24), or if the size of the personalization market shrinks (22 to 17). So factors outside of SDC's sphere of control are riskier if the Ad-Only option is pursued. If the likelihood of access to major clients is lessened, the prospects for the SDC weaken but TEMO is the favored approach (17 to 6). This highlights the dominant influence of Fisher Pennington in bringing SDC before major clients. If FP interest wanes, the prospects for SDC diminish somewhat but TEMO is still the more robust approach (25 to 23). So, factors closer to SDC sphere of control also favor TEMO. Directly within the realm of SDC's control are SDC's technical superiority, technical support, technical integration threat and client-service support. A confused message on technical superiority, poor technical support and client-service support, and a high threat in technological integration obviously diminish the prospects for SDC, but TEMO is still the more robust alternative (22 to 19). Why? SDC started with a segmentation scheme that was applicable to direct marketing. List scoring applications with major vendors such as Acxiom were available as a fallback. The basic understanding of the blocking and tackling of modern marketing, combined with the access to the state of the art in marketing science, provided a way the company could retrench and then move forward. As reported earlier, online retail sales grew 21% in 2001, with higher growth expected in 2002.[96] Email marketing grew 87% from 2000. Assisting in e-commerce transactions fills one of my online survival principles – stay close to the cash register. The further removed your service is from the point of transaction, the more likely it is to be viewed as a cost center rather than a revenue enhancer. These simulation results are summarized in Table 7.7 below.

Table 7.7. Comparative Valuations Under Different Scenarios.

What if?	SDC Valuation ($MM)	
	TEMO	Ad Only
Everything goes exactly according to plan	28	28
The growth in the e-commerce market flattens out	25	24
The Supply-Demand balance in Internet ads tips to oversupply	28	24
The size of the personalization market shrinks	22	17
The likelihood of access to major clients is lessened	17	6
FP interest wanes	25	23
There is a confused message on technical superiority, poor technical support and client-service support, and a high threat in technological integration	22	19

---

[96]*The State of Retailing Online 5.0: Performance Benchmark Report.* June 2002 Shop.org

Could we have known that advertising would go into the deepest downturn in 30 years? No. But we did know that advertising-industry expenditures are a mean-reverting process. Over the long haul, expenditures are expected to cycle back to their base rate of approximately 2.15% of U.S. GDP. Thus, the 12%-14% annual growth rate in aggregate U.S. advertising expenditures from Q1-99 through Q1-2000 was not sustainable. The downward trend in Q2-2000 (which continued to decline in Q3 and beyond) was recognizable as a return toward long-term, stable averages.[97] During the Q1-1999 through Q2-2000 period, the online population was expanding, unique reach increased from 23% to 28% of the U.S. population, and the time people spent online grew from 680 minutes per month per user to 980. During this same period, the Internet's share of U.S. advertising expenditure grew from 1.4% to 2.7%. While long-term pressures push advertising expenditure percentages toward the share of leisure time devoted to the Internet (share of eyeballs), shorter-term supply-demand balances had much more to do with the prices. These shorter-term imbalances came both from the collapsing of Internet advertising from venture-fund-supported Internet startups, and from the ease with which the surviving Internet publishing sites could increase their supply of advertising opportunities.

7.7   Strategic Planning Using the Four Risks

The strategic map for SDC developed using the path followed by all the early examples on the VentureDevelopmentProject.com Web site: Find the business kernel of the innovation, identify the appropriate first market, articulate the value proposition for that market, list the stakeholders, sketch the venture in value-network terms using Slywotzky's 11 dimensions, fill out the critical-issues grid, map the issues into a Bayesian network, and run the scenarios for best case, worst case, expected case, and other interesting cases.

The strength of this process is that it encourages deep strategic thinking (i.e., the simultaneous solution for the complex set of problems facing a venture). The weakness of this process is that moving from the critical-issues grid to a Bayesian network has been

---

[97] The statistics are from the Bureau of Economic Analysis, U.S. Department of Commerce.

an arduous task for several generations of students. So I have tried to find a prototype that would be easier to follow, and require less complete customization than the early examples. Using the four risks as an organizing theme for the Bayesian network seems to provide the needed prototype. What follows in this section is an example of building a strategic map for Core Micro Solution Systems using the four risks as an organizing principle.[98]

Remember that this kernel of the innovation is in the program control: using the fundamental properties of surface tension and adhesion at nano-scale to create programmable, discrete droplets that move independently over a dielectric surface. We tentatively identified the appropriate first market as high-throughput screening (HTS). The stakeholders (and market shares) in the overall sector were listed in Figure 6.1. HTS accounts for the major part of the biological screening and pharmacological testing budget for drug discovery. HTS is also becoming involved in toxicology and safety testing, and bioavailability.

Business Design

- *Fundamental assumptions:* The most prevalent problem in HTS is obtaining enough reagent in a timely manner to run the screening. Pharmaceutical and biotechnology companies are implementing strategies for miniaturization to counteract the difficulty in getting reagent. The benefits of miniaturization are low volumes requiring smaller amounts of reagents and less cost, with faster screening meaning more throughput. Overall, 80% of the laboratories plan to upgrade technologies to achieve the benefits of miniaturization in the next two years.
- *Customer selection:* CMSS should target pharmaceutical and biotechnology companies that use HTS techniques intensely and are willing to adopt innovative miniaturization technologies in order to achieve lower costs and faster results.

---

[98]This section is based in part on the work of the student team of Benjamin Chow, Peter Janda, Julie McDonald, Luciano Oliveira, Glenn Oyoung, and Arthur Wang, along with the discussions with Prof. CJ Kim, Wayne Lui and Patrick Deguzman, and the detailed input from Wayne Lui.

- *Scope:* CMSS should design and market biochips for biochemical assays, as microfluidics is believed to be a technology best suited for this kind of experiment. Manufacturing of the biochips can be outsourced, reducing the need for capital investments. Furthermore, the biochip uses standard materials easily available in the market and the manufacturing process employs standard machinery. There are companies that could perform production at a lower cost than a dedicated plant built only to meet CMSS needs.
- *Differentiation:* CMSS should pursue a differentiation strategy based on the kernel of its innovation: a programmable, standard platform to conduct experiments that allow pharmaceutical and biotechnology companies to perform faster assays using lesser amounts of reagents.
- *Value recapture:* The value is derived from savings in the cost of reagents, consumables, and automation. Robotics/liquid instruments, reagents and consumable costs are a major portion of laboratories' spending. The CMSS's EWOD technology can provide significant value in these areas. This value is monetized through the sale of disposable assay slides. The 480 labs in the high-throughput screening industry process some 2 million slides per year. Miniaturization could save companies substantial amounts of money in reagents and consumables.
- *Purchasing system:* CMSS should develop outsourcing contracts for manufacturing and strategic partnerships with automation and sensor companies for product development and distribution.
- *Manufacturing and operating system:* Manufacturing should be outsourced and managed in order to guarantee a high-quality product.
- *Capital intensity:* Substantial investment in product development will be needed in the early stage regarding the improvement of the biochips. Acceptable bounds for problems such as fouling, electrolysis and cross-contamination should be studied in order for clients to have a compelling reason to buy. Additional resources will have to be dedicated to the integration of the biochip with control systems and sensors.

- *R&D/Product development system:* CMSS should focus on product development as one of its core competencies.
- *Organizational configuration:* CMSS's organizational configuration should be traditional and very similar to most small, one-product start-up companies: centralized, pyramidal, and functional (R&D/Technology, Marketing & Sales, Finance/Administrative). Effective HR policies and incentives should be designed to attract excellent management and research professionals.
- *Go-to-market mechanism:* As CMSS's product will be complementary to those of other, more established players in the industry, it makes sense to share the sales force through a strategic partnership agreement.

The next step is the elaboration of the critical issues facing the venture, summarized in Table 7.8 below.

Table 7.8. Critical Issues Facing CMSS

	Company	Ecosystem	Infrastructure
Political	• UCLA License • Patent has not been approved yet	• Gov't/FDA requirements • Insurance policies	• Heightened security spending due to geopolitical events/trends • Different requirements for different countries
Behavioral	• Management talent • "Not engineered here" mentality • Whole product solution	• Competitor similar products entrenched, people familiar with them • Do end-users have to change their behavior to use our product?	• Companies want to go with large, well known companies (market leaders)

Economic	• Funding
	• Price/Value assessment
	• Is the business model viable based on market size, unit economics, returns requirement, etc?
	• Licensing
	• Scalability of production
		• Market size
		• Market ID
		• Funding
		• Life sciences
			• Abundant MEMS factory capacity
			• Labor market
			   o Short-term – abundant
			   o Long-term - questionable

Social	• Balanced management • Technology/Academic staff backgrounds	• Should we explore international opportunities?	• Graying of population
Technology	• Ability to serve many markets • Fouling • Longevity • Type of liquid	• Compatibility with other components within system • Nanolytics preemption	• Technological obsolescence

With these fundamentals we can proceed to a prototypical strategic map design, as shown in Figure 7.4 below. Four risk categories are the forces impacting the likelihood of "Venture Success," which in turn impacts the utility node "Venture Value." The decision node reflects the fundamental decision to launch the venture.

Figure 7.4. Prototype for a Strategic Map

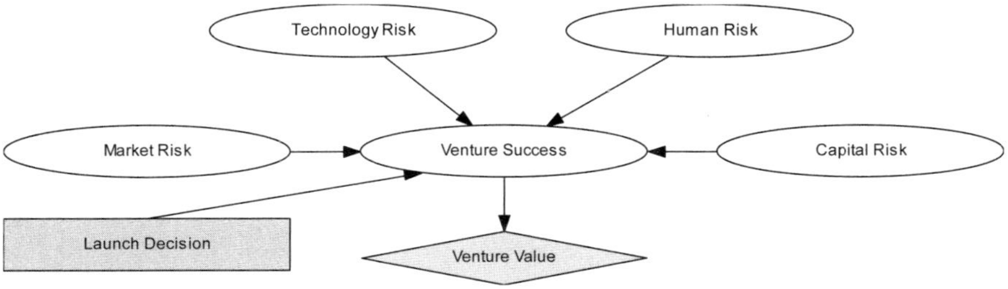

Addressing, one at a time, the factors that impact each of these risks provides a robust approach to creating a strategic map (Bayesian network).

Technology Risk

The technology risk that CMSS faces in pursuing the HTS market revolves around its
ability to adapt its current technology into a whole-product solution. To do so, CMSS must overcome four technical challenges related to its current technology: evaporation, compatibility with other systems, longevity, and fouling. Since CMSS must integrate into a whole-product solution, compatibility with other systems is a cost of doing business. But the other technical hurdles reflect decisions of investing none, a little, or a substantial amount (along with the consequent likelihood of solving the problem). That chunk of the map is shown in Figure 7.5.

Figure 7.5. The Factors Impacting Technology Risk

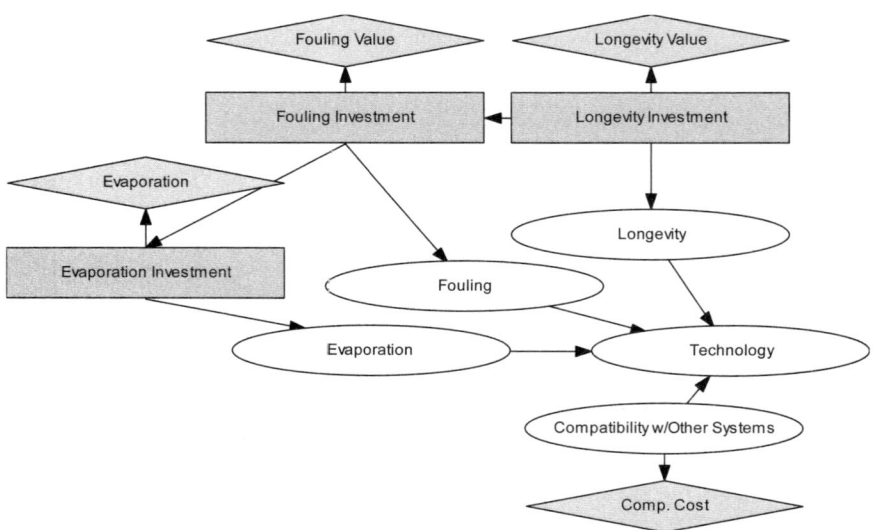

Market Risk

Four external factors determine CMSS's market risk in the HTS space: sales cycle length, go-to-market strategy, competitive threats, and market size/growth. Market size/growth also feeds directly into the venture value diamond, as seen in Figure 7.6.

Figure 7.6. The Factors Impacting Market Risk

Human Risk

The primary source of risk behind the human resources issues facing CMSS is having the right staffing mix and levels to: (a) overcome the technical challenges outlined above, and (b) manage the operations of CMSS from the business side and represent CMSS to outside parties. CMSS needs technical staff, management staff, and go-to-market staff.

The human risk node is modeled similarly to the technology node in that in each staffing area, CMSS has to decide the level of investment in human resources and the consequent likelihood that staffing will be adequate to meet the need. The sketch of these forces appears in Figure 7.7.

Figure 7.7. The Factors Impacting Human Risk

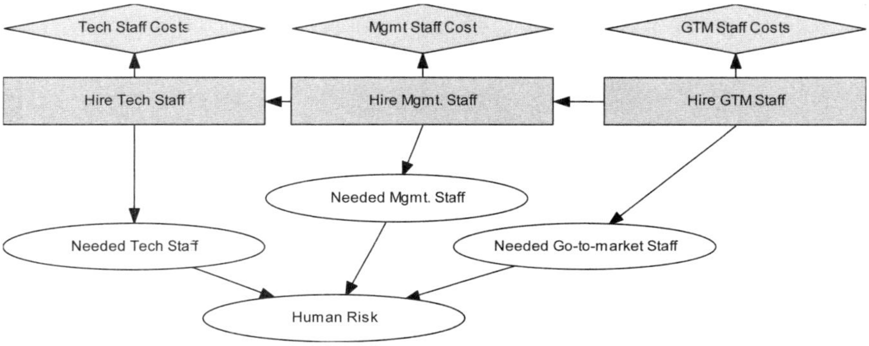

Capital Risk

The final risk facing CMSS is in the area of capital risk. CMSS's ability to achieve
venture success clearly depends on its ability to raise funds required to finance its growth. The capital risk node captures the influence of protectability of the intellectual property (IP), state of capital markets, expected time to exit, capital needed, and VC assessment of other risks. The VC decision has an obvious impact. The final component of the map appears in Figure 7.8.

234 | Midlife ~~Crisis~~ Startup

Figure 7.8. Factors Impacting Capital Risk.

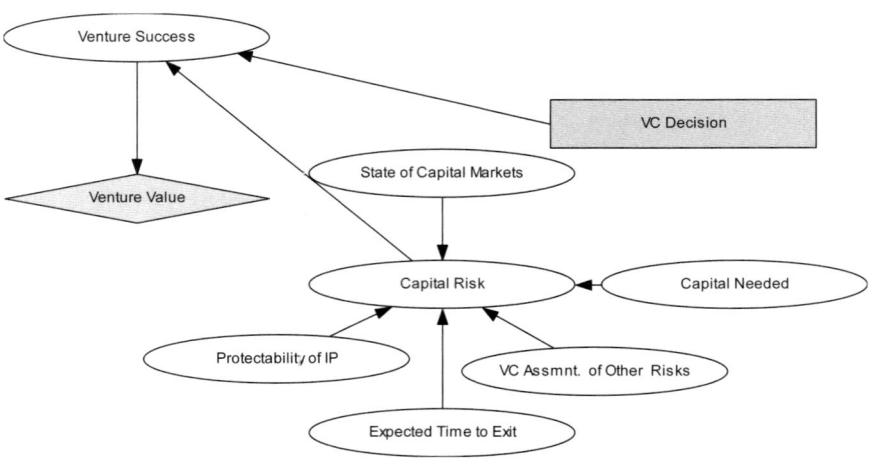

The resulting complete map appears in Figure 7.9. Note that the decisions have to be connected in time order. If timing is not critical, the time ordering will not have a major impact on the scenario results. But a time order must be established for the Bayesian network to compile.

Figure 7.9. The Complete Strategic Map for CMSS

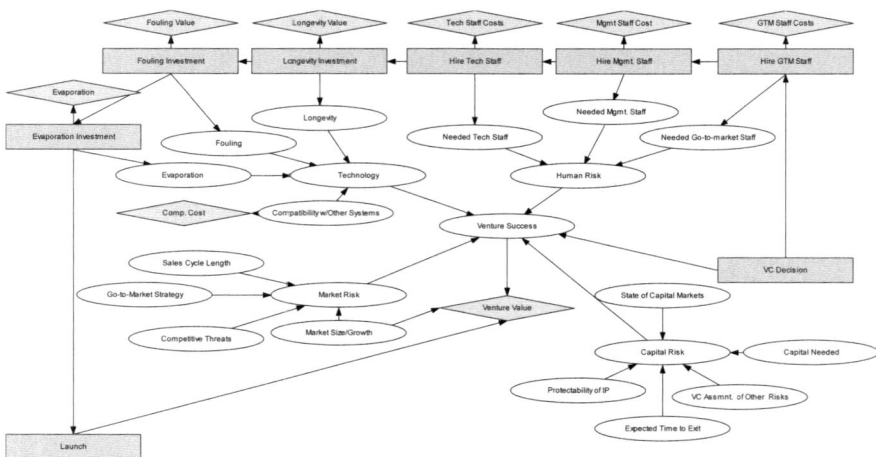

The student team and Wayne Liu patiently filled in all the required tables of likelihoods. The resulting network appears on the VentureDevelopmentProject.com Web site.

This map serves as a prototype for many technology ventures. The particular technology risks would vary by application, but the higher-order forces that impact the other risk nodes probably have a role in most ventures.

8. Meta Lessons

This final chapter addresses some of the overarching lessons from my new-venture experience. The first lesson concerns tenacity – an essential ingredient in a startup. The second lesson concerns sharing both a common picture of reality and fundamental values within a startup culture. The third issue concerns some of the barriers to bringing university technology to market. The fourth issue concerns the tensions that must be regulated in a startup environment and in life. The final lesson involves understanding the value of incremental progress. I end with a plan for the near-term future.

8.1 The Legend of Quincy Thomas

In *Good to Great*,[99] the prequel to *Built to Last*, Collins emphasizes the importance of getting the right people on the bus. When you do this successfully, *motivation* becomes a non-issue. You don't have to motivate the right people; they motivate themselves. This is Collins' way of addressing what I discussed in the section entitled "All Work is Voluntary." Using the value of entrepreneurial vision to help attract top people, I believe we did that. Once the brainpower is assembled it takes *tenacity* to bring results. The opening episode of this book ("Birth of a Notion") demonstrated the group's tenacity with immediate problems. The longer-term tenacity was tested only a few days later. Sunday, in the early afternoon, I received an email indicating our "baby monitor" had sounded an alarm. The program running the log files had quit for some unknown reason, but the automatic restart feature set things right without manual intervention. Then, at 1:10 p.m., that program stopped again, restarting automatically. When this happened again at 1:13, I began calling around. Neither Giovanni, Giuseppe, Ravi, nor Chuck answered their phones. I went down to the office, hoping someone was there working on it. No luck. I sat in the bullpen certain I could not deal with this alone. Finally I got through to Chuck on his cell phone, just

---

[99]Collins, James C. (2001), *Good to Great*, New York: HarperBusiness.

as he finished refereeing a youth basketball game. His alert had gone off, but he hadn't heard it during the game. As I related the story, he was as puzzled as I was. We decided to simply reboot the machine and hope for the best. Chuck walked me through the secure-shell communications with the Exodus servers, then through the rebooting of the VA Linux machines and the starting of test suite. We waited. Everything seemed fine.

This event became the major topic of the technology meeting Monday. The problem had not recurred, but we had no idea what the underlying issue was. No one wanted to let it go, but we had many pressing matters and I urged the group to move on. Several days later I walked into the second-floor offices and I saw Giuseppe smiling like the cat that ate the canary. "I figured it out!" he said. "It was Quincy Thomas." Our system gave a unique identifier to every registered iPlayer.com user who showed up on the few iPlayer.com pages we were using for our test. It consisted of the first three letters of the first name, the first letter of the last name, the five-digit ZIP code, and a one-character gender identification.

Have *you* figured it out yet? From the earliest parts of development of the test software, the programmers had put in a command to scan the program input for "quit" so they could stop a process if they ran into a problem. That command had never been removed. So Quincy Thomas showed up that Sunday early afternoon and the system stopped once it saw "QuiT" in the input stream. The incident speaks to a serious issue in software testing: It is never over. Unusual events signal a potentially deeper problem, even if they seem to go away. Other priorities may well take precedence, but a technology group that cares about the company never lets such issues die. Tenacity and pride in one's work play their parts,[100] but more is operating here. We had succeeded in establishing a culture that demanded everyone's best effort today and a constant emphasis on making it better tomorrow. While everyone bought into the entrepreneurial vision, the culture emphasized continually improving the company's products, rather than building a monument to the founders' vision. This is what Collins and Porras (1994) mean by "clock building," rather than "time telling," and by "good enough never is."

---

[100]For a classic treatment of the work culture of engineering teams, see Kidder, Tracy (1981), *The Soul of a New Machine*, New York: Little Brown and Company, republished in 1997 in a Modern Library Edition.

8.2 Sharing the Map

Having a mental map of a new venture in a form that can be shared is a tremendous asset to an entrepreneur. One lunch I was explaining my new-venture-planning approach to a colleague – a former venture capitalist now working in business development at UCLA. He asked me if companies really did this. I wanted to reply about how we were trying to improve the state of the practice when he basically answered the question himself. He said that as a VC, he insisted on a plan just to have a way of holding a CEO's feet to the fire. While he was thinking in terms of financial goals, the analogy goes much deeper.

The map in Figure 7.3 or 7.9 can be a shared picture of reality – the reality of the market and the economic web in which these companies operate. Having a vehicle that conveys the reality in a common form sets the stage for coming to a common sense of value among the key players – the board, the chairman, and the CEO. When founders are considering bringing in experienced management, the map can provide a vehicle for addressing hard-to-approach issues.

Long ago I was supposed to have learned that coming to a shared sense of reality was a prelude to discussing the important value issues that drive managerial decision-making. I used to teach this stuff. When we started UCLA's modern approach to management in the early 1970s, the first year was divided into two components: common-knowledge courses and the nucleus. The common-knowledge courses included basic accounting, economics, finance, statistics, human resources, and marketing. These were akin to the colors on an artist's palette – the building blocks of a managerial art. The creative art was designed into the more experiential components in the nucleus. The first-year nucleus consisted of three quarter courses: the first in individual decision-making, the second in managerial decision-making, and the third in managing complex systems. In the individual decision-making course – one of the courses I taught in the mid-1970s – we ended with a three-week series of games.

The first week, we used one of the many technology-based survival games. Teams were given a list of available materials and personnel, and a survival goal. Surviving a plane crash that left the group stranded in the desert, or in a lifeboat, or in remote Alaska, or on the moon, they had to move to a rescue location, taking only the most necessary equipment. While the exercise was really about some of the issues and traps in task-group communications, the MBA students invariably approached it by seeking some engineering or related expertise, thinking that technology had the right and wrong answers to the proper prioritization, and defending their problem solution as the right one in the face of equally viable, creative alternatives proposed by other teams. Students experientially learned that the agreement on a picture of reality helped them set the priorities and communicate their solutions.

Equipped with newly minted belief in their group decision-making skills, the students approached Exercise Kolomon.[101] This exercise presented a developing country with a relatively uneducated population, minimal infrastructure, substantial but undeveloped natural resources, and a potentially hostile set of geopolitical neighbors. Setting national priorities was the nominal group task: determining allocations to the military, the education sector, nascent industry, and infrastructure projects. In this exercise, students learned that establishing a common picture of reality was a necessary but not sufficient condition for problem solving. They struggled with intractable conflicts until they took the step back to ask about values. Only after seeking a consensus on the *value* issues involved could the group move toward a solution.

Now armed with task-group skills and aware of the need to share not only pictures of reality but basic values as a prelude to problem solving, MBA students entered the third week of exercises. The final exercise, called *Star Power*,[102] sets up a rigged game: a three-tiered, low mobility, hierarchical society in which arbitrary teams traded with other teams. Depending on the arbitrary assignment to a tier, the different teams started with varying levels of initial endowments – conveying a covert and unearned advantage that tended to persist in the trading game. After a number of trading rounds, the team that was ahead got to rewrite the trading rules for the next set of rounds. The sense of entitlement, justified or not, that went with success in the early rounds translated into a new set of rules that would make Machiavellians blush. The rules went from "You must agree to any

---

[101] I apologize to the authors, but I have no reference for this management game.
[102] The copy I have of the *Star Power* provides no insight into authorship. I hope someone is glad that the memory of the game has not been lost, even if the authorship has.

trade we demand" to "Give us all your coins." When I, as professor, would confront the winning teams with the obvious greed and short-term thinking inherent in the new rule set, the response, too often, was that I had unfairly tricked them. After they had written the most conspicuously unfair rules imaginable, I was the one who was being unfair. Maybe they were right.

There are two lessons here. The first concerns MBA programs and students, and the second concerns venture capitalists.

First, MBA students come with a sense of entitlement. They've succeeded in school, advanced in work, and gained entrance to prestigious MBA programs. Like Lake Woebegone, everyone is above average. I do not deny the accomplishments that have gone into MBA students getting to their current stage. I merely claim that they have to set aside that sense of entitlement to see the game clearly. Mine has not so far been the winning position. Student complaints over the immediate utility of the soft knowledge in process-oriented courses, along with the hegemony of the traditional disciplines, led to the nucleus being trimmed, a quarter at a time, out of existence.

The advocates for each color on the common-knowledge palette would correctly complain that students needed to know more about *red*, or students needed to know more about *blue*. The need for integrative experiences was replaced with courses in *the strategic uses of blue*, or *engineering applications of red*. The problem with MBA program design is analogous to what Collins and Porras (1994) call the "Tyranny of the OR." It was either more discipline-based courses OR more courses translating disciplinary knowledge into managerial art. The challenge, correspondingly, is to embrace the "Genius of the AND." How can we emphasize both disciplinary depth and trans-disciplinary artistry? While I do not claim to know how to resolve this dilemma for our MBA program in general, I do believe that new ventures pose fundamentally cross-disciplinary problems for managers. Immersing students in strategic problem solving for new ventures requires both disciplinary depth AND integration of skills.

One of the fundamental problems in tying technological lessons to their application is in recognizing how an ambiguous situation can be transformed into a structured one to which a known technology can be applied. In my early decades of teaching statistics to MBA students, I was continually disappointed at how they could easily answer the end-of-the-chapter exercises but fail essentially the same questions on a final exam. When they knew a problem called for a chapter-six solution, they were fine. When they had to determine which technique from their palette fit the situation best, the task was much more difficult. Trying to address the difficulties faced by new ventures forces MBA students to practice precisely the skills they will face most in their managerial careers. If the time devoted to MBA studies is going to add value to a career, then MBAs must figure out how to apply techniques they learned to somewhat ambiguous situations.

One fascinating example of the pedagogical potential comes from Professor James Theroux's "Real-Time Case" course at the University of Massachusetts, Amherst:

> As you read this, the managers of a new high-tech company, Optasite Inc., are striving to achieve the entrepreneurial dream. On a special password-protected Web site you will follow that company, and see their progress week by week. But you will do more than just watch. You will be actively engaged with the company, analyzing its problems, and making input. You will be participating in the first in-depth, real-time case study.
>
> Unlike traditional case studies, this real-time case will dig deeply into one company during an entire semester. At this moment, a case writer is stationed full-time at the case company. Each week the writer will provide us with the information we need to analyze a particular problem or question about the company. But our goal is not analysis for its own sake. Instead, we want to go beyond critiquing, and make valuable recommendations to the company. The company is counting on us to perform, and we want to deliver.[103]

Binding students to the ongoing stream of problems over the entire quarter or semester forces them to more richly assimilate the context of the case, understand how to abstract the critical issues, and more broadly employ their palette of skills. There are three tiers of information that are gathered for this real-time case. First is the

---

[103]Theroux, James (2002), "Supercharging the Case Method," presentation to the *Technology Enhanced Entrepreneurship Education (TE3) Clinic*, UmassOnline, October 10-12.

already-distilled management knowledge in the books and management journals relevant to the case issues. Second is the industry- or business-ecosystem-specific knowledge that provides much of the texture and context for analysis of managerial decisions. And last is the popular business press that records events that affect the planning contingencies. While Theroux took on the burden of assembling all three tiers, I believe responsibilities should be subdivided. Faculties should be responsible for the books and journal articles that address common issues across cases. These sources shouldn't get outdated in the time between course planning and execution. The company should provide the background on the industry or business ecosystem. And the students should be responsible for scanning the daily business press for relevant events in the political, behavioral, economic, sociological, and technological environments surrounding the company.

There is a second lesson from *Star Power*. I can't shake the eerie resemblance between the greedy behavior of MBA students in this exercise and the behavior of venture capitalists during the *chasm* and some CEOs when they replace founders. The chasm is the phase already described from Geoffrey Moore's technology adoption life cycle when the early enthusiasm wanes and potential mainstream customers take the show-me attitude of an economic buyer – requiring a whole-product solution and a compelling reason to buy. This is one of the emotional low points for the entrepreneur. The world no longer revolves around your vision or technological wizardry. You need just a little more time and money to respond to the pragmatic demands of the marketplace.

Given that the chasm is so well known and seemingly inevitable, perhaps I should not be surprised that venture capitalists write agreements that take advantage of it. While the agreements are struck during the early optimism, and *smart money* pays lip service to an alignment of interests, and partnering for the long term, clauses with full or partial ratchets turn major control over to venture funds and angels when the inevitable happens. True alignment of interests occurs when founders and funders share the same risks and opportunities. Pre-emptive rights could be maintained in a true alignment of interests, but all liquidation preferences would disappear, and investments in a venture would be at a current valuation. This would imply the need for only one class of stock, and simple shareholders agreements – far from the standard.

Venture capitalists often use the chasm as the occasion to wrest management control from the founders and bring in a CEO of their choosing. In my case it wasn't the chasm, but the press of commitments at UCLA that made me active in my search for a CEO. But once the reins were in his hands, he started rewriting the rules of the game. He was shielded from criticism for this by the *cult of the CEOs*. This is the cabal between venture capitalists and CEOs that too often dismisses technologically oriented entrepreneurs and founders as unrealistic dreamers who interfere with the real business. This is the tacit agreement on boards of directors that allows CEOs almost total freedom. The excesses that have led to arrests of high-flying CEOs and brought once-powerful companies into bankruptcy have started to erode the walls sustaining the cult. Much more needs to be done.

In particular CEOs of technology companies and VCs in technology arenas need to listen more closely to what the technological wizards have to say. Remember that radical innovation can impact many markets. The first market of application is just the one that is the low hanging fruit. Once rooted the technology can extend to other markets. Faculty members and other sources of innovation have a deep sense of the limits to which their technologies can be practically stretched. The technology pathway needs careful navigation. If the right questioned are asked and the answers are listened to, I believe innovations will have a much smoother path to their ultimate markets.

8.3   The University and Faculty Entrepreneurs

On the other side of this communications gap, universities and faculty entrepreneurs have much work to do to smooth the path of innovations to market. The Bayh-Dole Act[104] created a uniform patent policy among federal agencies that fund research, enabling small businesses and non-profit organizations, including universities, to retain title to inventions made under federally funded research programs.

---

[104]P.L. 96-517, Patent and Trademark Act Amendments of 1980, was co-sponsored by Senators Birch Bayh of Indiana and Robert Dole of Kansas and enacted on December 12, 1980.

Regarding universities, the major provisions of the act include:[105]
- Universities may elect to retain title to innovations developed under federally funded research programs;
- Universities are encouraged to collaborate with commercial concerns to promote the utilization of inventions arising from federal funding;
- Universities are expected to file patents on inventions they elect to own – trade secrets are not enfranchised as a method for developing intellectual property under this act;
- Universities are expected to give licensing preference to small businesses;
- The government retains a non-exclusive license to practice the patent throughout the world, and the government retains march-in rights that can lead to alternative licensing to those granted by universities;[106] and
- The Act encourages universities to participate in technology transfer.

The Bayh-Dole Act has had a major impact on universities. Prior to Bayh-Dole, fewer than 250 U.S. patents were issued to universities each year. In recent years, patents issued to U.S. universities have exceeded 2,000 per year. In FY 1999, technology transfer through licensing of innovations by U.S. universities, teaching hospitals, research institutes, and patent management firms added about $40 billion to the U.S. economy and supported 260,000 jobs.[107]

Given the mandate of and success attributable to the Bayh-Dole Act, how should faculty entrepreneurs be treated? The new regulations at UCLA describe three kinds of outside professional activities. The first kind includes taking an executive or managerial position in an

---

[105] This summary is based in part on material from the Association of University Technology Managers (see www.autm.net).
[106] A recent Op-Ed piece concerning using these march-in rights to restrain drug pricing produced a clarification from the authors of the Act. In "Our Law Helps Patients Get New Drugs Sooner," *The Washington Post*, Thursday, April 11, 2002, former Senators Bayh and Dole stated, "The ability of the government to revoke a license granted under the act is not contingent on the pricing of a resulting product or tied to the profitability of a company that has commercialized a product that results in part from government-funded research. The law instructs the government to revoke such licenses only when the private industry collaborator has not successfully commercialized the invention as a product."
[107] Association of University Technology Managers (see www.autm.net).

outside firm (possibly a faculty startup) or consulting contracts that extend beyond the normally allowed parameters (i.e., typically no more than a day a week for faculty outside the compensation schemes of the medical school). These now require prior approval delegated to the appropriate dean. The second kind includes board membership, including corporate boards and scientific advisory boards. This involves annual reporting and prior reporting when these entities are proposed as sponsors of research. Restrictions are possible due to conflicts of interests or conflicts commitment. The third kind involves editorial boards and normally permitted consulting. Such activities require no approval and are simply reported annually.

The *appropriate dean* is a troublesome phrase. At least three kinds of deans populate modern research universities. In public universities, one class of deans believes all science belongs in the public domain. The two decades since the passing of the Bayh-Dole Act that enfranchised private development of publicly funded research has done little to shake beliefs formed perhaps a half-century ago. To such deans, faculty entrepreneurs are tainted goods, possibly once-good researchers gone astray.

A second kind of dean, more typically found in professional schools such as business or law, understands the value of outside professional contact. Such contact is typically within the normal purview of consulting, which makes it easier for these deans to approach the relevant outside constituencies for discretionary funding. Sponsored research in these schools totals less than $1 million of UCLA's $655 million total for sponsored research in 2000-2001.[108] Beyond that purview, faculty startups present issues of possible conflict of commitment.

In professional schools such as engineering and medicine, the issue of sponsored research looms larger. The School of Medicine alone attracted more than $320 million in sponsored research for 2000-2001, and Engineering and Applied Sciences added $56 million.[109] A third kind of dean relies on the faculty's ability to attract large grants. The overhead on grants and the slice of patent royalties that goes to relevant departments provide incentives for these deans to take a more supportive view of the extension of a faculty's research agenda

---

[108]http://www.research.ucla.edu/report/fy00-01/awardbydept.htm .
[109]http://www.research.ucla.edu/report/fy00-01/awardbydept.htm .

toward practical or pressing national priorities. Grants provide annual revenues to these deans. Licensing of inventions provides a tangible revenue stream, while theoretically freeing the faculty to pursue more sponsored research – providing more overhead.

As previously described, my former dean was number two. I was not willing to seek his prior approval for my activity. My startup did not involve UCLA-developed technology, so it was easy to avoid a formal approval process. My stint as CEO ended before the new definitions of classes of outside professional activity became effective. I knew they were coming, and that was one reason I felt pressured into what turned out to be unfortunate decisions. Yet I believe my activity and the entrepreneurial activity of numerous faculty can be synergistic with these faculty's other university roles and commitments. My needs, and the needs of many other faculty entrepreneurs, would be better served if the university's basic stance allowed faculty to maintain roles with decision-making authority in startups, so that they could more easily determine the fate of their innovations. Worry more about actual conflicts of interest or commitment, and less about prior restraint on faculty options.

In a number of ways, most faculty are entrepreneurs. One of the few real freedoms in *academic freedom* is over the research agenda a faculty chooses. But to pursue any agenda, faculty must procure the needed resources. Even major research universities provide very little infrastructure for unsponsored research: a library, a computer-network infrastructure, perhaps a computer tied to that network, an office, and hopefully stimulating colleagues with whom to share ideas. Some faculty members need little more than a pad of paper to demonstrate the "superior intellectual achievement" that is the fundamental requirement for promotion.[110] But for faculty whose chosen research agenda falls into medical, biomedical, engineering, or

---

[110]For purposes of advancement and promotion, the performance of faculty members is evaluated by grouping their activities into four interrelated categories: teaching, research and creative work activity; professional competence and activity; and university and public service. Of these, teaching and scholarly or creative activity clearly are primary activities and receive the largest commitment of effort and energy, but faculty members are also expected to participate in university activities and to contribute to their professions and to the community. "Superior intellectual attainment, as evidenced both in teaching and in research or other creative achievement, is an indispensable qualification for appointment or promotion to tenure positions." Adapted from *The UCLA Call: A Summary of Academic Personnel Policies and Procedures*, http://www.apo.ucla.edu/call/.

many other arenas, major funding is required for laboratories, equipment, and to support staff and apprentice personnel. The entrepreneurial task is to gather the human and financial resources that are required to pursue one's chosen agenda. Academic freedom is a hollow ideal if one can choose to inquire only where the resources are intense – like the drunk looking for his lost keys under the lamppost because that's where the light is. So the faculty entrepreneur finds the ways to bring light to the desired area of inquiry.

Sometimes the desired area of inquiry aligns with already articulated industry goals, such as in the six research partnerships of the University of California's Industry-University Cooperative Research Program (IUCRP): BioStar (biotechnology), CoRe (communications and networking research), DiMI (digital media innovation), LS:IT (life sciences & information technology), MICRO (microelectronics innovation), and SMART (electronics manufacturing). Sponsoring companies provide at least a one-to-one match of the system-wide funds requested. Confidential peer review, contract review, and conflict-of-interest review make these grant opportunities very akin to the processes in National Institutes of Health (NIH) grants. The industry sponsorship provides additional resources, helps ensure aligned agendas, and identifies candidates for possible commercial licensing of any innovations that result from the research. For example, a California biotech company figures out how to extract stem cells from adult skin and muscle, and how to induce these cells to differentiate into neuronal cells. A UCLA researcher interested in Parkinson's disease wants to figure out how to further induce these cells to differentiate into the kinds of cells destroyed by Parkinson's. The alignment of agendas is obvious. Licensing of resulting intellectual property is open to all bidders to avoid any conspicuous pipelining of research findings to a favored company. If a materials-transfer agreement were needed for this research, then the issues might become more complex. Otherwise, this is an easy example of how aligned and articulated research interests can lead to a natural hand-off of academic research into commercial application to further worthwhile goals.[111]

But commercial applications do not appear as Athena – springing fully grown out of the head of Zeus. The licensing company may

---

[111] Chesselet, Marie Francoise, "Pluripotent Stem Cells and Parkinson's Disease," BioStar Grant Number: 01-10195.

want to consult with Professor Zeus after the research is complete. Professor Zeus may want to stay involved in the further development of the Parkinson's treatment, because it is part of his or her chosen research agenda. At this point, the university begins to create barriers that can inhibit academic freedom, as well as commercial opportunity. In UCLA's case, the Office of Intellectual Property Administration (OIPA) places itself as the arbiter of IP rights even in the case of consulting that uses no university resources.

> Intellectual property rights: Companies normally would like to have all patents and other intellectual property assigned to them as a condition of the consulting arrangement. It may be possible to make such assignments when the work is done without the use of University resources, in the company's facilities and outside the scope of the faculty member's primary University responsibilities. An assignment of patent rights, however, may only be made after disclosing any new invention to the University. OIPA and the Office of Technology Transfer will determine the degree to which the University will assert rights to any new invention based on the circumstances of the invention's derivation.[112]

This sets up potentially adversarial relations among faculty, the university, and the outside company. We are back to the scope of work debate from Chapter 2, and the exceptions to the patent agreement that faculty sign embedded in the details of the California Labor Code:

> This agreement does not apply to an invention which qualifies under the provisions of Labor Code section 2870 of the State of California which provides that (a) Any provisions in an employment agreement which provides that an employee shall assign, or offer to assign, any of his or her own time without using the employer's equipment, supplies, facilities, or trade secret information except for those inventions that either: (1) Relate at the time of conception or reduction to practice of the invention to the employer's business, or actual or demonstrably anticipated research or development of the employer. (2) Result from any work performed by the employee for the employer. (b) To the extent a provision in an employment agreement purports to

---

[112] http://www.research.ucla.edu/oipa/facultyconsultagrmts.htm .

> require an employee to assign an invention otherwise excluded from being required to be assigned under subdivision (a), the provision is against the public policy of this state and is unenforceable.[113]

OIPA has taken initiative recently to aid faculty understanding of the startup process. A half-day seminar in the spring of 2002 provided a panel of experienced venture capitalists, bankers, lawyers, and accountants who discussed how to distinguish which projects in a university lab might form the basis for starting a company, and a panel of faculty entrepreneurs who presented lessons learned from their experience starting companies based on their research. In the fall of 2002, OIPA and the Tech Coast Angels (TCA) established a new "Investor Forum" where faculty will have an opportunity to meet with TCA members to discuss start-up projects. The first session in the spring seminar was on the rules of the game and emphasized how disclosing ideas or inventions to the university is the ante for beginning the process. The screening questionnaire for the TCA Investor Forum first asks about disclosure, then follows with a litany of details:

- Two-sentence project description
- Applicant background: Domain expertise and research background.
- Do you have collaborators on this project from UCLA, other institutions, or companies?
- How has the technology development been funded to date? By whom?
- What is your solution to the problem? What is the market opportunity for your technology?
- Why is your technology unique and what are the competitive advantages?
- Who will be your first customer and what is your revenue model for your product/service?
- Have you published or publicly presented this technology?[114]

Veteran inventors perhaps know what invention disclosure means in their field, and may not be surprised at the information requested prior to discussion with TCA. But the process is daunting to first-

---

[113]Quoted from the UC Patent Agreement:
http://www.ucop.edu/ucophome/policies/bfb/g40a.html.
[114]Adapted from: http://www.research.ucla.edu/oipa/officehours/.

time faculty entrepreneurs. This asks more than they are likely to know, seemingly making faculty responsible for areas that TCA should be bringing to the interchange. The form may call for disclosure before the ideas are ready for that level of formalization. Premature formalization may result in the loss of the commercialization opportunity. Most likely, faculty have heard cautionary tales of being ensnarled in the turgid university bureaucracy, and have no idea of the time commitment implied by disclosing something that may or may not have commercial potential. Turning the judgment over to the university when the university does not have a great track record in commercializing faculty IP implies a loss of control most faculty members would rather avoid. Best practices from the point of view of the university technology licensing office (TLO) may not be very comforting from the faculty perspective. Allan (2001) reports the experience at Yale in which 1% (10 of 850) of total disclosures led to 70% of licensing revenues, and 88% (748 of 850) of disclosures generated less than $10,000 each, Yale's approximate cost for processing an invention disclosure.[115] The statistics at UCLA are even more extreme, since a single, 1989 patent for treating intercranial aneurisms generated over 74% of the $8.3 million revenue for 2000-2001.[116] Obtaining early disclosure, providing rapid screening, and focusing time and energy on those disclosures most likely to succeed financially is a sound recipe for efficient operation of a TLO, but not one that is likely to encourage faculty to interact with the bureaucracy.

Efficiency for TLOs or outside agencies such as the TCA pushes these organizations toward skimming the cream off the top – rapidly finding the most likely commercial successes rather than supporting the research agenda of the faculty serving the broader needs of the community for economic development, or serving the broader societal need for remedies and innovations. Efficiency for faculty (at least for those outside the medical device and pharmacological arenas where clinical trials dominate the product-development process) involves publishing the basic science and forming outside companies (or aligning with existing outside companies) that cite the published

---

[115] Allan, Michael F. (2001), "A Review of Best Practices in University Technology Licensing Offices," *The Journal of the Association of University Technology Managers*, Volume XIII, 57-69. Also see the Yale Office of Cooperative Research home page, in particular, Senior Director Gregory E. Gardiner's 1997–98 report at http://www.yale.edu/ocr/images/docs/ocr_report_96-98.pdf.
[116] UC Technology Transfer Annual Report 2001.

science as prior art while patenting the specific extension of the basic work. These differential approaches to efficiency establish a potentially adversarial relationship between the university and its faculty, and fail to properly engage the university's broader role in research and service.

I believe the innovation potential of the university should be mined, not skimmed. TLOs are invariably placed under the auspices of a vice chancellor for research. I believe, consequently, some attention should be paid to how this process serves the research agenda of faculty.

My research agenda for many years has involved developing the models and methods for guiding managerial decision making in information-rich environments. Understanding choice when the naturally occurring record of consumer choice overwhelms most managers' intuitive capacities, or building models of competitive market structure from the empirical record of market transactions, are the kinds of challenges I enjoy. I did it in the 1980s and 1990s for scanner data from retail transactions in mass-market outlets. Then I needed only data, which came for free from commercial sources, and the computational infrastructure that the Anderson School provided. I believed that same resource mix would not suffice for building the infrastructure for marketing science in the Internet Age. I would need much more computing power, much more manpower, and data structured differently than the browser logs that had started to dribble out of some Internet firms. Interacting with Internet firms to get good data was frustrating. They were operating at an extremely fast pace. Data were accumulating faster than storage capacity to retain it. Vendors were taking 36 hours to process 24 hours worth of session logs. A number of startups I interacted with in my faculty role were throwing out what I considered to be vital records of their early interactions with customers. Standard grant arrangements were unlikely to buy me the tools to redress these problems. So my BHAG (big, hairy, audacious goal),[117] was to build a company, Strategic Decision Corp., that would supply the manpower, computer power, and data (action logs) designed to my specification. Then the broader community of marketing scientists would have an easier time developing the models and methods appropriate to this new age. I

---

[117] See Collins and Porras (1994).

didn't trust the university to be a partner in this effort, and at least the academic part of this dream was lost as a partial consequence.

I do not believe my mistrust was misplaced. In my ideal university, things would be different, but I think my judgment about UCLA at the time was accurate. The question then becomes, how can the university embrace such entrepreneurial activity?

Carnegie Mellon University (CMU) has made some very progressive steps in this direction.[118]
It has articulated the vision that:

> "Carnegie Mellon should encourage the creation of innovations on campus and then facilitate the timely and effective transfer of those innovations to the outside community. When commercialization would be the most effective mechanism for this transfer, the university needs to have policies, procedures and services in place ... to ensure that the transfer proceeds smoothly and without unreasonable barriers or delays. The new approach recommended here positions the university as facilitator rather than adversary. It draws upon and supports the university's collaborative, entrepreneurial culture. It puts stronger and more productive campus-wide and regional connections into place."[119]

Carnegie Mellon's traditional version of a TLO was started in the early 1990s under the assumption that it should become a substantial profit center for the university. "Operations were structured so that staff attention would be allocated in billable hours to dockets deemed to have the greatest likelihood of creating large royalty streams or capital gains."[120] The spinoff of Lycos accounted for more than $40 million of the $61 million generated from tech transfer between 1993 and 2000. Despite the one hit (or maybe because of it), CMU's University Research Council felt the "AUTM-driven emphasis on financial indicators of success" focused the tech-transfer role on revenue generation rather than CMU's "... commitment to the

---

[118]This material was adapted from Emerson, S. Thomas (2002), "Carnegie Mellon: Innovation Exchange," presentation to the National Consortium of Entrepreneurship Centers Directors Conference, Babson College October 4.
[119]Carnegie Mellon University, University Research Council (2001), "Recommendations from Deliberations of 2000-2001 Academic Year," November 4.
[120]Ibid, p. 6.

creation and transfer of innovations as integral to the university's mission in service to society."[121]

This basic shift in thinking enfranchised a wonderfully different approach. CMU created the Carnegie Mellon Innovation Network to provide active assistance in the innovation-transfer process to inexperienced innovations needs, along with simplified and streamlined processes for experienced innovators, active facilitation of connections with CMU alumni and the region's resources for innovators, space for incubation, the possibility that faculty could hold management positions in startups concurrent with their university appointments, faculty committees as an optional alternative to relying on the discretion of department heads (or deans) in deciding on appropriate faculty roles in startups, education programs to stimulate entrepreneurship, and an innovation culture on campus including: "education and training in the basics of commercialization; courses that bring an entrepreneur's perspective into the classroom; faculty-to-faculty courses to stimulate interdisciplinary collaboration; courses that focus on the "how-to's" of developing a commercial product, bridging the tension between promotion and production, and product innovation; expanded entrepreneurship programs for graduate and undergraduate non-business majors and regional entrepreneurs; and entrepreneurship-focused campus events."[122]

CMU also encourages benchmarking of its efforts compared to other universities. I hope the university succeeds and leads the way for others.

8.4   The Regulating Tension of Opposites

I see the world of faculty and business as talking past each other. Although their fates are inextricably tied, not enough attention is paid to managing the tensions between them.

I learned something about such tensions long ago. At the very end of the 1970s I took a sabbatical in Greece. I was leaving an administrative position as director of UCLA's Arts Management Program and returning to my research on perceptual mapping and choice (market-share) modeling. I spent most of the time Mykonos in a house we rented on a bluff overlooking the harbor and the island of

---
[121] Ibid, p. 11.
[122] Ibid, p. 14.

Delos. I had no phone to interrupt my thinking. In addition to my primary academic agenda, I had formulated two side tasks. The first was to find a coin that granted admission to the Festival of Dionysus in 5th Century BCE Greece. I never succeeded in that. The second was to find the meaning of *enantiodromia*, a perplexing term I ran across while reading Jung. Jung had defined it as "a running contrariwise" and attributed it to Heraclitus. Since I knew I'd be visiting the American School for Classical Studies in Athens, I decided to find the original Greek term in Heraclitus – hoping some clarification would come from that.

Heraclitus was considered the last of the Ionian school of material monists – philosophers who believed that the world was made up of one kind of stuff and that that stuff was matter. Mind and soul were looked on as matter.[123] He was referred to as "Heraclitus the Dark" by the ancient scholars who could read all his work, most of which was burned in the fire of the great library of Alexandria. "... (T)o the moderns, who possess only isolated sentences, he is darker still."[124] The good side of having so little of his writings remain is that the task of searching for *enantiodromia* was shortened. While I didn't find that term in Greek, I did find the term *palintonos* (the regulating tension of opposites) at the core of his philosophy:

> LVI. The attunement of the world is of opposite tensions, as is that of the harp and the bow.[125]

This thought is juxtaposed with a series of related fragments:

> I. It is wise to listen, not to me but to the Word, and to confess that all things are one.
> LVII. Good and bad are the same.
> LXIX. The road up and the road down is one and the same.
> XLV. They understand not how that which is at variance with itself agrees with itself.
> XXXV. The teacher of most men is Hesiod. They think he knew many things, though he did not understand day and night. For they are one.

---

[123]Heraclitus, *On the Universe*, with an English translation by W.H.S. Jones, London: William Heineman Ltd., New York: G.P. Putnam's Sons, MCMXXXI, from the introduction.
[124]Ibid, p. 453.
[125]Ibid.

"The most characteristic difficulty in Heraclitus philosophy lies in the demand which it places on its adherers to transcend the 'either-or' type thinking and to recognize in each phrase the existence that a relationship of 'both-and' may be present."[126]

I do not believe it is difficult to recognize that "good and bad" are two sides of the same coin, just as day and night are two sides of the same orb. But "two sides of the same coin" makes one think of "heads or tails," of "either or," which is not the point. The essence is more that "good and bad" are two extremes of the same dimension. To drive home the point, consider the definition of a single concept in science. Even the narrowest definition of a concept in science (i.e., the explicit definitional form) asserts that a concept is synonymous with the set of operations used to measure it. Think about a simple semantic differential scale that calls on a respondent to rate something on a continuum that goes from *good* to *bad*.

Good   1   2   3   4   5   Bad (circle a number).

Whether you call it two ends of the same continuum or two sides of the same coin, *good* and *bad*, viewed this way, embody the same scientific concept. That part is easy epistemology of science. Where Heraclitus transcends into other philosophical realms is in the assertion that there is a regulating tension between these two extremes (or any such two extremes). To play a tune on this lyre of life you must regulate the tensions between these opposing forces. With too little tension, no note emerges. To increase the tension a constant increment takes ever-increasing amounts of energy. The more extreme you are, the exponentially more effort it takes to become yet more extreme. If one stretches too much in one direction the tension, of course, may snap. Jung uses this notion to indicate how your *shadow* (i.e., the opposite of your dominant personality) can snap out of nowhere and kick you in the butt if you become too extreme in your dominant personality constellation.

I had then, and now, little interest in the personality dynamics Jung considered. Rather, some of this thinking reinforced the kinds of market-share models I was developing at the time. A simple

---

[126]Wheelwright, Philip (1959), *Heraclitus*, Princeton: Princeton University Press, p. 91.

translation of *palintonos* is "elastic."[127] The principle of *palintonos* should be manifest in market-share elasticities. One would expect that a brand's share elasticity approaches zero as the share for that brand approaches one. That is, as one's market share becomes more extreme, it takes ever-increasing amounts of underlying expenditures to increase share more. That property differentiates the attraction models I worked on to reflect market share and choice from the linear and multiplicative market-share models that were traditional alternatives.[128] The regulating tension of opposites was also fundamental to my notion of distinctiveness. Marketing actions must be distinctive to be effective in a market-share sense. Yet to be ever more distinctive requires exponentially increasing effort.[129]

This philosophy has two uses in the current context. In an abstract sense, it should help readers understand what Collins and Porras (1994) mean by the "Tyranny of the OR" versus the "Genius of the AND." By understanding and regulating the tension between united but opposing forces, Collins and Porras urge companies to seek:

purpose beyond profit	AND	pragmatic pursuit of profit
a relatively fixed core ideology	AND	vigorous change and movement
...		
Big Hairy Audacious Goals	AND	incremental evolutionary progress
...		
ideological control	AND	operational autonomy[130]

Second, the regulating tension of opposites is severed when venture capitalists come in and toss out the original entrepreneurs. This is so accepted within the VC community, and yet so antithetical to what

---

[127]Feyerabend, Karl (undated), *Langenscheidt's Pocket Greek Dictionary: Greek-English*, Berlin and Munich: Langenscheidt KG.
[128]Cooper, Lee G., and Masao Nakanishi (1988), *Market Share Analysis: Evaluating Competitive Marketing Effectiveness*, Boston: Kluwer Academic Publishers, Chapter 2. An online edition is available at:
http://www.VentureDevelopmentProject.com/MCI_Book/new_page_2.htm.
[129]Ibid. Chapter 3.
[130]Collins, James C. and Jerry I. Porras (1994), *Built to Last: Successful Habits of Visionary Companies*, New York: HarperBusiness, p. 44.

Collins and Porras found leads to long-term success. But Collins and Porras also emphasize that it is visionary companies, not visionary leaders, who help build lasting greatness. Entrepreneurs and venture capitalists need to recognize that building companies that last takes contributions from both. Expect ongoing tension, but use that tension to create a sympathetic vibration in the environment that surrounds the joint endeavor.

8.5 *A Place to Begin and a Path to Make It Better*

When I teach strategic marketing planning, I emphasized that the mental maps can be very subjective at first, and improve over time. As we learn more about a particular relationship, or the impact of legislation or other events, we can do the revisions and elaborations that seem appropriate. Step by step we can turn the heuristic mental map into the quantitative decision-support tool it was designed to be. The current alternatives are a normative, speculative exercise based on game theory, constructive generalizations built on agent-based simulations, or the rhetorical exercise of relating scenarios. The core idea of strategic maps is to transform the underpinning of management strategy into more of an empirical, quantitative science – a foundation that naturally incorporates incremental improvements. Change the assumptions in a normative model, and you need to re-derive all the results and implications. Change the first principles in an agent-based simulation and you must re-run all the simulations, and re-characterize the emergent behavior. Change the path followed in a scenario and the rhetorical tale could be vastly different.

Having a system that allows for incremental improvements is fundamentally important. Radical innovations are akin to long leaps over rugged landscapes, but once the innovation hits ground it must root itself incrementally into the stream of nutrients that will sustain it, or it will die. The notion of continual incremental improvement underlies the work of W. Edwards Deming. It's fundamental to Petroski's (1992) record of how useful products evolve.[131] It's behind the Collins and Porras (1994) principles of "preserve the core/stimulate progress," "try a lot of stuff and keep what works," and "good enough never is." Collins (2001) iterates the principle in "the flywheel."

---

[131] Petroski, Henry (1992), *The Evolution of Useful Things*, New York: Alfred A. Knopf.

Mostly what we have to incrementally improve is the way these worlds communicate with each other. In particular the business world has to learn how to listen to the technological world regarding the limits of technology and the technological world needs to understand and communicate better how the kernel of its innovation can be better mapped into the world of application.

8.6   *What's Next?*

I've stressed the importance of having a place to begin and a path to make it better. So what's the next step in this path for me? I had no clear idea when I began writing this. I know that without the infrastructure that SDC was supposed to provide, I will not be doing methods development on Internet data. I also know it is a mistake to define one's future by what one cannot or will not do. Attraction to interesting areas, rather than avoidance of problematic ones, is a better path for me.

As my writing has proceeded, the clearest problem that evolved was how the centers of innovation within the university and the masters of the business world talk past each other. Since major business schools stand with feet in both worlds, we should be able to do something about this. At UCLA we are going to try. UCLA's huge research engine is cranking out innovations at great speed, but very little commercialization of those innovations is occurring. The business development specialists at UCLA's Office of Intellectual Property Administration (OIPA) help bring faculty researchers in contact with VCs, can help identify what needs to be done, but don't have the manpower to help them do it. The Anderson School's MBA students, on the other hand, are very interested in learning about new-venture initiation, but are not organized in a way that systematically helps UCLA faculty bring innovations to market. Four-person teams or individual "independent studies" are the standard. These provide very little manpower and no knowledge accumulates from one team's effort to the next.

I've used the thinking that went into the writing of this book to design a course entitled "Strategic Marketing Planning for New Ventures." That course matches teams of Anderson students with faculty innovators. We want our students to be the clinical manpower – learning about the new-venture creation by building the market assessments and business plans that faculty innovators need to take their next steps toward commercialization. We intend to develop knowledge management systems that can grow and be updated – helping the next generation of efforts learn from the past.

Within the Price Center for Entrepreneurial Studies in UCLA's Anderson School we are revitalizing the Venture Development Project (VDP). I've been appointed the faculty director of that project. We plan to develop the infrastructure to support my new class, two existing courses in business plan development and venture initiation, the venture-initiation option within our two-quarter capstone Applied Management Research course, and the independent studies that have historically been the most isolated efforts.

It has been quite a ride. I propose a toast to the challenges ahead – one taught to me by Giovanni from his Sicilian roots that captures the spirit of courage needed to confront the inherent uncertainties of a new venture. "In bocca al lupo." "In the mouth of the wolf."

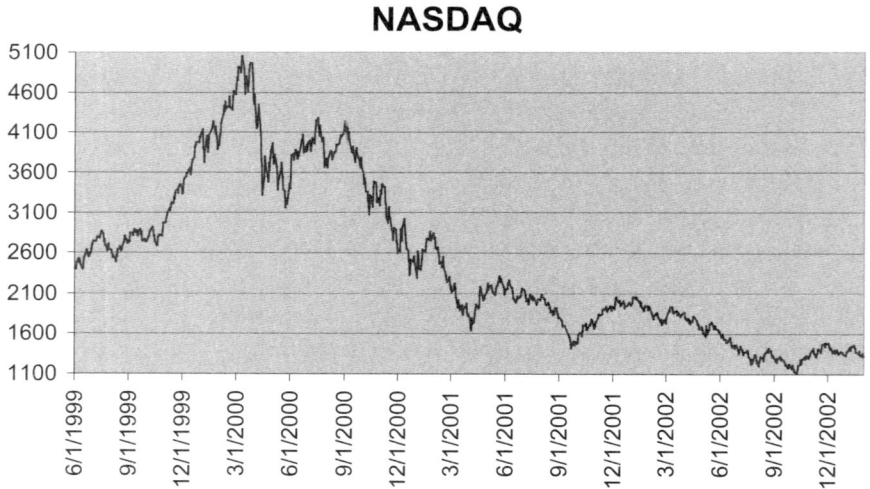

References

Allan, Michael F. (2001), "A Review of Best Practices in University Technology Licensing Offices," *The Journal of the Association of University Technology Managers*, Volume XIII, 57-69.

Bayh, Birch, and Robert Dole (2002), "Our Law Helps Patients Get New Drugs Sooner," *The Washington Post*, Thursday, April 11.

Belasco, James A. (1990), *Teaching Elephants to Dance: The Manager's Guide to Empowering Change*, New York: Crown.

Belasco, James A. and Ralph C. Stayer (1993), *Flight of the Buffalo: Soaring to Excellence, Learning to Let Employees Lead*, New York: Warner Books, Inc.

Belasco, James A. and Jerre Stead (1999), *Soaring with the Phoenix: Renewing the Vision, Reviving the Spirit, and Recreating the Success of Your Company*, New York: Warner Books, Inc.

Bower, Joseph L. and Clayton M. Christensen (1995), "Disruptive Technologies: Catching the Wave," *Harvard Business Review*, January-February, 44-53.

Brooks, Frederick P. Jr. (1995), *The Mythical Man-Month: Essays on Software Engineering Anniversary Edition*, Reading Mass: Addison Wesley Longman, Inc.

Carnegie Mellon University, University Research Council (2001), "Recommendations from Deliberations of 2000-2001 Academic Year," November 4.

Cialdini, Robert B. and David Schroeder (1976), "Increasing Compliance by Legitimizing Paltry Contributions: When Even a Penny Helps," *Journal of Personality and Social Psychology*, 34:599-604.

Cialdini, Robert B., John T. Cacioppo, R. Bassett and J. Miller (1978), "Low-Ball Procedure for Producing Compliance: Commitment Then Cost," *Journal of Personality and Social Psychology*, 36:463-476.

Chesselet, Marie Francoise, "Pluripotent Stem Cells and Parkinson's Disease," BioStar Grant Number: 01-10195.

Christensen, Clayton M. (1997), *The Innovator's Dilemma: When New Technologies Cause Great Firms to Fail*, Boston: Harvard Business School Press.

Coase, Ronald H. (1937/1952), "The Nature of the Firm," in *Readings in Price Theory*, George. J. Stigler and Kenneth. E. Boulding, eds. Chicago: Irwin, 331-51.

Collins, James C. and Jerry I. Porras (1994), *Built to Last: Successful Habits of Visionary Companies*, New York: HarperBusiness.

Collins, James C. (2001), *Good to Great*, New York: HarperBusiness.

Cooper, Lee G. (1981), "Some Perspectives on Art, Organizational Behavior and Democracy," *The Journal of Management and Law of the Arts*, 11, 1-26.

Cooper Lee G. (2000), "Strategic Marketing Planning for Radically New Products," *Journal of Marketing*, 64, 1 (January), 1-16.

Cooper, Lee G. (2003), "Personalization and Technology-Enabled Marketing," in Kamarkar, Uday and Uday Apte (editors), *Current Research on Managing in the Information Economy*, Boston: Kluwer Academic Publishers, in press.

Cooper, Lee G., Penny Baron, Wayne Levy, Michael Swisher, and Paris Gogos (1999), "PromoCast: A New Forecasting Method for Promotion Planning," *Marketing Science*, 18, 3, 301-316.

Cooper, Lee G. and Giovanni Giuffrida (2000), "Turning Datamining into a Management Science Tool," *Management Science*, 46, 2 (February), 249-264.

Cooper, Lee G., and Masao Nakanishi (1988), *Market Share Analysis: Evaluating Competitive Marketing Effectiveness*, Boston: Kluwer Academic Publishers.

Cooper, Lee G., Masao Nakanishi, and Dominique M. Hanssens (1982), "Estimating Cross Competitive Influences on Market Share," The Procter and Gamble Invitational Psychometric Conference, October.

Cooper, Lee G., Troy Noble, and Elizabeth Korb (1999), "Strategic Marketing Planning in Turbulent Environments: the Case of PromoCast," *Canadian Journal of Marketing Research*, 18, 46-66.

CRM ASP Opportunities, Dataquest, August 23, 1999.

Culbert, Samuel A. and John B. Ullmen (2001), *Don't Kill the Bosses! Escaping the Hierarchy Trap*, San Francisco: Berrett-Koehler.

Darwin, Charles (1859), *On the Origin of Species by Means of Natural Selection*, p. 210.

Dickens, Charles (1850/1948), *David Copperfield*, Garden City, N.Y.: Literary Guild of America.

Emerson, S. Thomas (2002), "Carnegie Mellon: Innovation Exchange,"
presentation to the National Consortium of Entrepreneurship Centers Directors Conference, Babson College October 4.

Emery, Fred E. and Eric L. Trist (1965), "The Causal Texture of Organizational Environments," *Human Relations*, 18, 1, 21–32.

Ferguson, Charles H. (1999), *High Stakes, No Prisoners: A Winner's Tale of Greed and Glory in the Internet Wars*, New York: Times Business – Random House.

Foster, Richard (1986), *Innovation: The Attacker's Advantage*, New York: Simon & Schuster.

Fox, Sandra (2002a), "Fine-tuning the Technology Strategies for Lead Finding," *Drug Discovery World*, Summer.

Fox, Sandra (2002b), "High-throughput Screening 2002-New Strategies and Technologies," High Tech Business Decisions, Inc.

Ghiselli, E.E., J.P. Campbell, and S. Zedeck (1981), *Measurement Theory for the Behavioral Sciences*, New York: Freeman.

Giuffrida, Giovanni, Wesley W. Chu, and Dominique M. Hanssens (2000), "Mining Classification Rules from Datasets with Large Number of Many-Valued Attributes." *Proc. 12th Int'l Conf. on Extending Database Technologies (EDBT)*, Konstanz, Germany, March.

Giuffrida, Giovanni, Lee G. Cooper, and Wesley W. Chu (1998), "A Scalable Bottom-Up Data Mining Algorithm for Relational Databases," in *10th International Conference on Scientific and Statistical Database Management* (SSDBM '98), Capri, Italy, July, IEEE (Institute of Electrical and Electronics Engineers), Publisher.

Godin, Seth, and Don Peppers (1999), *Permission Marketing: Turning Strangers into Friends, and Friends into Customers*, New York: Simon & Schuster.

Grove, Andrew S. (1996), *Only the Paranoid Survive: How to Exploit the Crisis Points That Challenge Every Company*, New York: Doubleday.

Guilford, J.P. (1967), *The Nature of Human Intelligence*, New York: McGraw-Hill.

Haeckel, Stephan H. (1999), *Adaptive Enterprise: Creating and Leading Sense-and-Respond Organizations*, Boston: HBS Press, p.17.

Hagel, John III (1996), "Spider Versus Spider," *The McKinsey Quarterly*, 1.

Hagel, John III and Arthur G. Armstrong (1997), *Net Gain: Expanding Markets Through Virtual Communities*, Boston: Harvard Business School Publishing.

Hagel, John III and Marc Singer (1998), *Net Worth: Shaping Markets When Customers Make the Rules*, Boston: Harvard Business School Publishing.

Jupiter Communications (1999), "Proactive Personalization: Learning to Swim, Not Drown in Consumer Data," August.

Jupiter Vision Report, "Online Advertising Through 2005: Flourishing in the Dot-com Decline," August, 2000.

Kano, Noriaki, Shinichi Tsuji, Nobuhiko Seraku and Fumio Takahashi (1984), "Miryokuteki Hinshitsu to Atarimae Hinshitsu (Attractive Quality and Must-be Quality)," *Journal of Japanese Society for Quality Control*, 14, 2.

Kaplan, Jerry (1995), *Startup: A Silicon Valley Adventure*, Boston: Houghton Mifflin.

Kauffman, Stuart E. (1988), "The Evolution of Economic Webs," in *The Economy as an Evolving Complex System, SFI Studies in the Sciences of Complexity*, Philip W. Anderson, Kenneth J. Arrow, and David Pines eds., Reading, MA: Addison-Wesley Publishing Company.

Kauffman, Stuart E. (1995), *At Home in the Universe: The Search for Laws of Self-Organization and Complexity*, Oxford: Oxford University Press

Kidder, Tracy (1981), *The Soul of a New Machine*, New York: Little Brown and Company, re-published in 1997 in a Modern Library Edition.

Koestler, Arthur (1941), *Darkness at Noon*, New York: The Macmillan Company.

Kornelis, Marcel (2002), "Modeling Advertising Markets Using Time-Series Data," doctoral dissertation, Rikjsuniversiteit, Groningen, The Netherlands, p. 115.

MacFarquhar, Larissa (2002), "Whom Do You Call When an Executive is Unbearable?" *New Yorker*, April 22 & 29, 114-136.

Maslow, Abraham (1943), "A Theory of Human Motivation," *Psychological Review*, 50, 370-396.

Maslow, Abraham (1954), *Motivation and Personality*, New York: Harper.

Massarik, Fred (1985), "Human Experience, Phenomenology, and the Process of Deep Sharing," in Tannenbaum, Robert, Newton Margulies, Fred Massarik and Associates (eds.), *Human Systems Development*, New York: Jossey-Bass, 26-41.

Mena, Jesus (1999), *Data Mining Your Web Site*, New York: Digital Press.

Moore, Geoffrey A. (1991), *Crossing the Chasm: Marketing and Selling Technology Products to Mainstream Customers*, New York: Harper Business.

Moore, Geoffrey A. (1995), *Inside the Tornado: Marketing Strategies from Silicon Valley's Cutting Edge*, New York: Harper Business.

Moore, Geoffrey (2000), *Living on the Fault Line: Managing for Shareholder Value in the Age of the Internet.* New York: Harper Business.

Moore, James F. (1996), *The Death of Competition: Leadership & Strategy in the Age of Business Ecosystems*, New York: Harper Business.

Pearl, Judea (1986), "Fusion, Propagation and Structure in Bayesian Networks," Cognitive Systems Laboratory Technical Report CSD-850022 R-42, Department of Computer Science, UCLA (revised March 1986).

Pearl, Judea (2000), *Causality: Models, Reasoning, and Inference*, Cambridge, UK: Cambridge University Press.

Perloff, Harvey S., Paul Bullock, Lee G. Cooper, Simon Eisner, and Hyman R. Faine (1979), *Arts in the Economic Life of the City*, New York: American Council for the Arts.

Petroski, Henry (1992), *The Evolution of Useful Things*, New York: Alfred A. Knopf.

Porter, Michael (1980), *Competitive Strategy*, New York: The Free Press.

Porter, Michael (1985), *Competitive Advantage*, New York: The Free Press.

Robertson, Thomas S. and Hubert Gatignon (1998), "Technology Development Mode: A Transaction Cost Conceptualization," *Strategic Management Journal*, 19, 6, 515-531.

Schoenberger, Chana R. (2002), "Marketing: Web? What Web?" Forbes Online, June 10, 2002.

Schwartz, Peter (1996) *The Art of the Long View*, New York: Currency Doubleday.

Scott, Carol A. and Richard F. Yalch (1980), "Consumer Response to Initial Product Trial: A Bayesian Analysis," *Journal of Consumer Research*, 7 (June), 32-41.

Shane, Scott, and Toby Stuart (2002), "Organizational Endowments and Performance of University Start-ups," *Management Science*, 48, 1 (January), 154-170.

Shapiro, Carl and Hal R. Varian (1999), *Information Rules: A Strategic Guide to the Network Economy*, Boston: Harvard Business School Press.

Shop.org (2002), *The State of Retailing Online 5.0: Performance Benchmark Report*, June.

Slywotzky, Adrian J. (1996), *Value Migration: How to Think Several Moves Ahead of the Competition*, Boston: Harvard Business School Press.

Swire, Peter P. and Robert E. Litan (1998), *None of Your Business: World Data Flows, Electronic Commerce, and the European Privacy Directive*, Boston: The Brookings Institute.

Thaler, Richard (1999), "Mental Accounting Matters," *Journal of Behavioral Decision Making*, 12, 183-206.

Theroux, James (2002), "Supercharging the Case Method," presentation to the *Technology Enhanced Entrepreneurship Education (TE3) Clinic*, UmassOnline, October 10-12.

Tversky, Amos and Daniel Kahneman (1981), "The Framing of Decisions and the Psychology of Choice," *Science*, 211, 453-458.

von Bertalanffy, Ludwig and Anatol Rapoport (1956), *General Systems: Yearbook of the Society for the Advancement of General Systems Theory, Volume 1*, Ann Arbor, MI: Society for General Systems Research.

West, Patricia, Dan Ariely, Steve Bellman, Eric Bradlow, Joel Huber, Eric Johnson, Barbara Kahn, John Little, and David Schkade (1999), "Agents to the Rescue," *Marketing Letters*, 10, 3 (August), 285-300.

Wind, Yoram (1978), "Issues and Advances in Segmentation Research," *Journal of Marketing Research*, XV (August), 317-333, for an introduction to the special issue.

Wolff, Michael (1998), *Burn Rate: How I Survived the Gold Rush Years on the Internet*, Simon & Schuster.

Index

academic freedom 247, 248, 249
ad optimization 6, 69
alignment ... 10, 27, 47, 48, 243, 248
Allan, Michael F 251, 261
Ariely, Dan 268
Atari 13, 29, 38, 152, 153, 154, 155, 157, 158, 163
Bayesian networks 42, 214, 225, 227, 228, 231, 234
Bayh-Dole Act ... 244, 245, 246, 261
Belasco, James A. 117, 261
Bellman, Steve 268
best practices 124, 127, 130, 131
Bower, Joseph L. 16, 261
Bradlow, Eric 73, 84, 268
bridge financing 91, 92, 114, 119, 134, 143, 145
Brooks, Frederick P. Jr. 50, 96, 261
burn rate 46, 73, 91, 98, 146
business constraints 80, 81, 92, 93
business development 5, 29, 37, 47, 49, 50, 160, 168, 239, 259
business ecosystem 14, 17, 20, 126, 194, 200, 206, 207, 209, 243
business plan ..... ix, 25, 28, 34, 43, 48, 55, 57, 104, 151, 166, 167, 169, 171, 180, 187, 189, 190, 259, 260
business-to-business networks 70
Capital Risk 28, 233, 234
Chesselet, Marie Francoise 248, 262
Christensen, Clayton M. 16, 17, 80, 90, 130, 152, 156, 160, 261, 262
Chu, Wesley W. 31, 264
Cialdini, Robert B. 97, 261, 262
Coase, Ronald H. 18, 262
code escrow 68, 206
Collins, James 165, 237, 238, 241, 252, 257, 258, 262
competitive market structure 252
conditional tables 216, 217, 225
conflict of commitment 51, 52, 53, 246
conflict of interest 103, 104, 111, 119, 136, 247, 248
Cooper, Lee .. 8, 11, 22, 30, 51, 177, 186, 213, 257, 262, 263, 264, 266
corporate culture ... 42, 82, 165, 209, 237, 238, 253, 254

Critical Issues Grid...... 22, 195, 199, 227, 230, 242
cross-functional teams..........90
Culbert, Samuel A. ....x, 90, 263
Darwin, Charles....14, 198, 263
datamining..... 8, 24, 25, 26, 28, 31, 32, 34, 35, 36, 38, 39, 40, 43, 48, 49, 56, 79, 125, 168, 170, 171, 178, 179, 188, 190, 191, 197, 198, 201
decision making.... 4, 23, 29, 41, 48, 160, 178, 239, 240, 247, 252
delighters 89
Dickens, Charles............ 44, 263
Digital F/X 29, 75, 77, 154, 157, 181
dilution....... 38, 102, 103, 106, 107, 111, 113, 146
disruptive technologies.... 16, 17, 20, 80, 130, 156, 159, 163, 261
due diligence 43, 59, 74, 78, 88, 111, 129, 151, 189, 190
Emerson, S. Thomas 253, 263
Emery, Fred E. 17, 19, 263
End of Life..............................20
entrepreneurial vision ... 14, 96, 97, 151, 164, 165, 166, 237, 238
exit strategy 146, 166
feature creep............................89
Ferguson, Charles H. 61, 73, 74, 99, 263
Financial Risk.........................28
first market....... 22, 151, 156, 159, 161, 227, 228, 244
Foster, Richard 16, 17, 263
Fox, Sandra ....x, xi, 159, 162, 263, 264
Friday Surprises 80, 81, 88

full ratchet 116, 118, 243
Gatignon, Hubert.......... 18, 267
Ghiselli, E.E................. 137, 264
Giuffrida, Giovanni 8, 30, 31, 262, 264
Godin, Seth 123, 212, 264
go-to-market strategy.... 31, 34, 49, 50, 91, 130, 168, 230, 232
Grove, Andrew S. 17, 264
Guilford, J.P.................. 44, 264
Haeckel, Stephan H. 164, 165, 264
Hagel, John III 19, 70, 208, 264
Hanssens, Dominique M. ...30, 31, 70, 263, 264
Heraclitus 255, 256
Huber, Joel...........................268
Human Risk 27, 28, 55, 95, 206, 232, 233
hygiene features.....................89
incremental improvements 17, 126, 130, 258, 259
intellectual property 30, 37, 38, 48, 51, 52, 53, 54, 56, 122, 126, 152, 154, 169, 171, 189, 191, 192, 200, 201, 203, 204, 205, 212, 214, 218, 233, 244, 245, 246, 248, 249, 251
investor groups...111, 112, 113
Johnson, Eric......................268
Jupiter Communications 72, 170, 171, 176, 178, 182, 183, 265
Kahn, Barbara....................268
Kahneman, Daniel 84, 267
Kano, Noriaki............... 89, 265
Kaplan, Jerry 32, 73, 265
Kauffman, Stuart E........ 14, 19, 265

kernel of the innovation 17, 22, 151, 152, 153, 154, 155, 156, 157, 158, 159, 160, 163, 227, 228, 229, 259
kickback 104, 105
Kidder, Tracy 238, 265
Kiretsu 136, 137, 138, 208
Kleiner Perkins 28, 29, 75, 76, 157
knowledge management 33, 56, 260
Koestler, Arthur 133, 265
Korb, Elizabeth 22, 213, 263
Kornelis, Marcel 123, 213, 265
lift 68, 69, 72, 73, 80, 114, 118, 120, 126, 129, 131, 133, 134
linear satisfiers 89
Litan, Robert E. 204, 267
Little, John D. C. 268
living-systems theory 14, 15, 16, 19, 22
MacFarquhar, Larissa 117, 265
Main Street 20, 21
market finding 17, 126, 130, 152, 156, 160
Market Risk 28, 232
marketing science 6, 8, 30, 33, 39, 51, 84, 102, 122, 126, 128, 129, 137, 143, 164, 170, 209, 212, 215, 219, 226, 252
Maslow, Abraham 96, 265
Massarik, Fred 45, 266
Mena, Jesus 266
Moore, Geoffrey 16, 19, 23, 31, 33, 37, 152, 180, 243, 266
Moore, James F. 16, 266
multi-channel marketing ... 122, 124, 125, 198, 209, 211, 225

must haves 89
Nakanishi 51, 89, 186, 257, 263
Noble, Troy 22, 41, 87, 200, 213, 263
Office of Intellectual Property Administration 54, 249, 250, 259
organizational boundaries ... 14, 22, 29
Pearl, Judea 68, 214, 266
Peppers, Don 123, 212, 264
Perloff, Harvey S. 11, 266
PersonalClerk 6, 56, 57, 66, 73, 80, 86, 91, 98, 124, 125, 127, 169, 170, 171, 172, 173, 174, 175, 178, 179, 180, 181, 182, 183, 187, 188, 206, 209, 210
personalization .... ix, 5, 6, 26, 32, 33, 37, 38, 44, 48, 56, 70, 121, 122, 123, 124, 126, 138, 164, 170, 171, 172, 176, 177, 178, 179, 183, 196, 197, 198, 204, 205, 206, 207, 211, 212, 215, 218, 222, 223, 225, 226
Petroski, Henry 258, 266
Porter, Michael ... 193, 194, 266
positioning. 11, 21, 34, 40, 165, 196
pre-emptive rights 111, 112, 113, 145, 146, 147, 243
pre-money valuation ... 73, 101, 107, 108, 112, 119
privacy 5, 25, 32, 108, 110, 122, 126, 170, 173, 176, 198, 204, 205, 212, 215, 218, 219
PRIZM 40
product development ... 42, 48, 57, 89, 90, 91, 93, 109, 115, 206, 229, 230, 251

PromoCast 8, 22, 30, 39, 41, 213, 262, 263
radical innovation 4, 16, 23, 151, 152, 160, 244, 258
Rapoport, Anatol 15, 268
recommendation engine 6, 24, 40, 49, 66, 68, 69, 73, 79, 91, 108, 109, 124, 196, 201
referenceable accounts 40, 107, 110, 122, 208, 212
retirement 83, 84, 140
risk tolerance 137
Robertson, Thomas S. .. 18, 267
scenario planning 214, 219
Schkade, David 268
Schoenberger, Chana R. 207, 267
Schwartz, Peter 214, 267
scope of work .. 53, 54, 55, 249
Scott, Carol A. 97, 267
SEC 29, 74, 75, 76
segmentation 5, 11, 20, 21, 24, 40, 56, 57, 67, 68, 69, 71, 80, 92, 95, 107, 109, 124, 165, 167, 168, 171, 172, 173, 174, 176, 178, 179, 183, 188, 190, 198, 201, 209, 226
Series A 37, 46, 50, 51, 55, 101, 102, 103, 105, 111, 113, 189
Series B 5, 49, 74, 75, 77, 101, 102, 105, 111, 112, 113, 116, 119, 167, 189, 213, 225
Series C 92, 116, 118, 119, 120, 121, 130
Series D 75, 134, 135, 138, 144
Series E 143, 144
Shane, Scott 27, 267
Shapiro, Carl 18, 267
Singer, Marc 70, 264

Slywotzky, Adrian 34, 43, 158, 167, 193, 227, 267
spaghetti code 94
stakeholders 22, 195, 227, 228
strategic map 22, 121, 227, 228, 230, 231, 258
strategic marketing planning .... x, xi, 11, 14, 22, 27, 126, 135, 140, 193, 213, 258, 259
supply chain 23, 41, 161
sustainable competitive advantage 20, 151
Swire, Peter P. 204, 267
Tannenbaum, Robert ... 45, 266
targeting 5, 11, 24, 40, 68, 124, 165, 169, 171, 172, 173, 174, 175, 176, 177, 178, 179, 197
Technology Risk 28, 55, 231, 232
technology-enabled marketing .. ix, 5, 11, 24, 28, 36, 40, 45, 101, 122, 123, 124, 126, 127, 136, 137, 164, 167, 193, 196, 204, 210, 211, 212, 213, 222, 223, 224, 225, 226
terms of engagement 104
Thaler, Richard 84, 267
The Bowling Alley .. 20, 21, 31, 167, 180
The Chasm 19, 20, 171, 243, 244
The Tornado 20, 21
Theroux, James 242, 243, 267
transaction costs 18
Trist, Eric L. 17, 19, 263
turbulent environment 19, 22, 165, 194
Tversky, Amos 84, 267

UCLA..... 7, 10, 13, 26, 29, 32, 41, 43, 50, 51, 52, 53, 54, 62, 70, 76, 82, 83, 84, 93, 95, 99, 136, 142, 158, 165, 206, 209, 239, 244, 245, 246, 247, 248, 249, 250, 251, 253, 254, 259, 260

valuation 34, 35, 36, 63, 78, 83, 86, 91, 92, 101, 102, 106, 107, 108, 109, 110, 112, 114, 115, 118, 119, 122, 138, 140, 145, 196, 210, 211, 213, 225, 226, 243

value proposition..... 17, 18, 22, 28, 33, 35, 127, 128, 158, 161, 162, 163, 167, 171, 173, 190, 191, 197, 227

value recapture.....................180

value-based pricing 21, 48, 68, 147, 168, 180, 182

Varian, Hal R. 18, 267

vesting....................77, 102, 140

virtuous cycle 157

von Bertalanffy, Ludwig...... 15, 268

West, Patricia 268

whole-product solution...... 15, 18, 19, 20, 21, 36, 161, 163, 169, 206, 231, 243

Wind, Yoram 268

Wolff, Michael........73, 74, 268

ZipSegments 40, 48, 49, 50, 53, 56, 67, 71, 102, 107, 122, 124, 127, 144, 171, 174, 175, 178, 180, 188, 209, 211